# An Aid to the International Code of Botanical Nomenclature

A.N. HENRY
**Regional Botanist**

*and*

M. CHANDRABOSE
**Botanist**
BOTANICAL SURVEY OF INDIA, COIMBATORE

**TODAY AND TOMORROW'S PRINTERS AND PUBLISHERS**

*First Published 1980*

*Price* **Rs. 30.00**
**$ 6.00**

*Distributed by*

**ISBN : 0-88065-104-0**

**SCHOLARLY PUBLICATIONS**
7310 El Cresta Drive
HOUSTON, Texas 77083 (USA)

*Composed by* : Sanjay Composing Agency, New Delhi-110005.
*Printed at* : Mohan Printing Corporation, Delhi-110007

*Published by :* Today & Tomorrow's Printers and Publishers
24-B/5 Desh Bandhu Gupta Road, Karol Bagh, New Delhi-110005.

# Foreword

PLANT nomenclature, an integral part of taxonomic botany, is an advance subject of biological significance and international importance. It deals with the procedure to be followed in determining the correct botanical name to be applied to a plant after it has been identified and classified. In order to understand the scope and *modus operandi* of plant nomenclature a study of the provisions in the International Code of Botanical Nomenclature is essential. Several of the nomenclatural problems are very complex indeed ! This book while dealing with the various provisions of the Code has made the study of technically intricate and legalistic situations easier by providing annotated examples. To the writing of this book the authors have brought their scientific training in research taxonomy and practical experience. I am confident that this volume will be of considerable use to all those interested in finding solutions to nomenclatural problems of Indian plants. Also, from the point of view of those Indian University teachers who are not professional taxonomists, and students who are eager to learn the history, development and present status of plant nomenclature, this book is certainly very welcome.

COIMBATORE
15th May, 1979

N. C. NAIR
*M. Sc., Ph.D., D.Sc., F.B.S.,*
*Deputy Director,*
*Botanical Survey of India,*
*Southern Circle,*
*Coimbatore-641002*

# Preface

IN writing this book it has been our object to provide a simplified and annotated version of the "International Code of Botanical Nomenclature" with suitable illustrative examples from the Indian plants, so as to benefit all research workers in Botany, including teachers and students of Indian Universities for a better understanding of the various provisions of the Code and also for their practical application. The official Code in its legislative and esoteric language, is a document daunting when used without special training or guidance. The correct interpretation of the Articles in particular, is found difficult unless the reader/user is thoroughly familiar with the specialized language of nomenclature. For many years we have had the opportunity to discuss with various research workers including Professors and Students in Botany their problems concerning the study of Plant Nomenclature, and this book is largely to cater to their needs; simple language is used in the book to interpret the various Articles in a collective way with adequate explanation of the context in which they are defined.

The paramount importance of the correct identity and nomenclature of plants of all groups needs no emphasis. It is hoped that this book, which provides a practical guide to a monotonous and complex field of plant nomenclature, will serve as a useful tool in the hands of all plant lovers to promote understanding of the principles and practice of Botanical Nomenclature, thereby enabling them to fix correctly the names of plants.

Our grateful thanks are due to Prof. Dr. K.K. Lakshmanan, Professor and Head, Department of Botany, University of Madras Autonomous Post-graduate Centre, Coimbatore for his helpful suggestions in framing the contents of the book according to the current syllabi of various Universities in India; also to Mrs. C. K. Vijayam for neatly typing the manuscript in time.

COIMBATORE
7 May, 1979

A.N. HENRY AND M. CHANDRABOSE

# Contents


FOREWORD ____ III

PREFACE ____ V

CHAPTER

1. INTRODUCTION ... ... ... 1
2. HISTORY OF THE CODE OF BOTANICAL NOMENCLATURE ... ... ... 3
3. POLYNOMIAL AND BINOMIAL SYSTEMS ... 10
4. BROAD OUTLINE OF LENINGRAD CODE (1978) 12
5. MAIN DIVISIONS OF THE INTERNATIONAL CODE OF BOTANICAL NOMENCLATURE (1978) ... 13
6. RANKS AND NOMENCLATURE OF TAXA ... 15
7. DEFINITIONS ... ... ... ... 18
8. THE TYPE METHOD OR TYPIFICATION ... 23
9. GUIDE FOR THE DETERMINATION OF TYPES ... 31
10. PRIORITY ... ... ... ... 33
11. EFFECTIVE AND VALID PUBLICATION ... 35
12. CITATION OF AUTHORS' NAMES AND OF LITERATURE AND GENERAL RECOMMENDATIONS ON CITATION ... ... ... ... 38
13. RETENTION OF NAMES OR EPITHETS OF TAXA ... 43
14. CHOICE OF NAMES ... ... ... 44
15. REJECTION OF NAMES AND EPITHETS ... ... 46
16. ORTHOGRAPHY OF NAMES AND EPITHETS ... 48
17. HOW TO DESCRIBE A NEW GENUS/SPECIES ... 50
18. NOMINA CONSERVANDA ... ... ... 61
19. SOME IMPORTANT DATES IN BOTANICAL NOMENCLATURE ... ... ... 64
20. INTERNATIONAL ASSOCIATION FOR PLANT TAXONOMY (IAPT) ... ... ... 67
21. TEACHING OF BOTANICAL NOMENCLATURE THROUGH PRACTICAL APPLICATION ... ... 70
22. EXERCISE ... ... ... ... 78
23. LITERATURE ... ... ... ... 89
24. GLOSSARY CUM INDEX ... ... ... 91

# 1

## INTRODUCTION

Plants have always been useful to man, and plant nomenclature arose because there was need for names which could serve as vehicles of communication. There is no precise record of the beginnings of plant nomenclature for it is obscure as the records of earliest civilizations. In Sanskrit nomenclature, the genus is indicated by a name while the species is described by a prefix denoting its distinctive character: BALA is the modern genus *Sida*, and MAHA-BALA=*Sida rhomboidea*, ATI-BALA=*Sida rhombifolia*, and NAGA-BALA=*Sida spinosa*. Modern Botany originated in Europe in the 16th Century A.D. Since the various languages of different countries could not furnish precise names for all plants, botanists all over the world were compelled to invent a vocabulary of their own for consistency and uniformity in scientific names. Latin vocabulary was universally accepted because of the following :

(i) Latin was the common universal language of Scholars.

(ii) Latin is specific and exact in its meaning which is particularly pertinent to the needs of descriptive phrases of the biological sciences.

(iii) Most of the old literature that deal with plants are either in Greek or Latin.

Further an acceptable system of nomenclature must be international in character and, therefore, the scientific names of plants although derived from any language must be latinized. The binomial system of nomenclature was consistently used by Linnaeus in his *Species Plantarum* (1753), and this date has been accepted as the starting point for present day botanical nomenclature.

A set of internationally accepted rules was felt imperative for the stabilization of scientific names of plants. The International Code of Botanical Nomenclature, as known in recent times, is derived mainly from the Laws of Botanical Nomenclature proposed by Alphonse de Candolle in 1867. These laws, in their turn, are mainly based on the various aphorisms and pronouncements clearly stated by Linnaeus in his *Fundamenta Botanica* (1736) and explained in great detail in *Critica Botanica* (1737). The text of the current edition of the *International Code of Botanical Nomenclature* (Stafleu *et al.* 1978) is based upon the decisions reached by the Nomenclature Section of the 12th International Botanical Congress held in Leningrad from 3rd to 10th July 1975. It is the product of the intense study by the best brains in the botanical fraternity, who for nearly a century has been studying the problems connected with the naming of plants. In various botanical congresses held generally at an interval of about five years, every effort was taken to make the system satisfactory in all respects and to secure a stable and uniform system of plant nomenclature by way of suitable amendments to the Code.

# 2

## HISTORY OF THE CODE OF BOTANICAL NOMENCLATURE

Linnaeus in his *Fundamenta Botanica* (1736) and *Critica Botanica* (1737) gave various aphorisms and pronouncements for naming plants. Later, in *Philosophia Botanica* (1751), he amplified the views expressed in his aphorisms and principles, thereby establishing a real beginning of a sound system of plant nomenclature. It may be mentioned here that in the seventeenth and eighteenth centuries, there were no generally accepted rules for naming of plants, except the Linnean Principles, and it was difficult to realise the confusion which existed over plant names; every Botanist followed his own way in naming the plants. According to the Linnean Principles, no two genera could have the same generic name and no two species within a genus could have the same specific epithet. If a genus was divided into two or more genera, the original name of the genus was retained for one among them. Priority of publication was given recognition, in general. The binomial system arose from Linnaeus, *Species Plantarum* (1753) wherein he used consistently the so called trivial name (specific epithet) towards the margin of the text against each species. He employed this system to avoid assigning different names to a species or identical names to different species.

Since plants new to science increased so rapidly confusion in nomenclature existed. There was no index of plant names from which botainsts of different countries could learn the names of plants that had been used. In 1813 Augustin de Candolle published *Theories elementaire dela botanique* containing instructions on various nomenclatural procedures, mostly followed by Linnaeus. Steudel published his

*Nomenclator botanicus* (1821), indexing the names of flowering plants with their synonyms.

**Paris Code (1867)**

The first organised efforts towards standardization and legislation of nomenclatural practices were taken at the Ist International Botanical Congress in Paris in August 1867. Alphonse de Candolle proposed his "*Lois de la nomenclature botanique*" mainly based on his father's (Augustin de Candolle) *Theorie elementaire de la botanique* (1813) which in turn was derived mainly from the various aphorisms and pronouncements stated by Linnaeus. The "laws of botanical nomenclature" which formed the "Paris Code" was printed and published in 1867. It consisted of only three chapters with a total of 68 articles. The rule of priority was treated as fundamental. Considerable attention was given to author citations and to the terms applied to various categories of plants. The requirements for vaild publication and for acceptance and rejection of names were established. The rules of the Paris Code represented an excellent beginning in the right direction, but their application revealed many deficiencies.

**Kew-Rule**

This rule which is no longer legislated by the Code, was essentially to retain the first epithet given to a species in a certain combination. In other words the earliest specifiic epithet was not retained in changing the generic name of a plant. The Kew-Rule which had major divergences from the Paris Code later led to much controversy especially between the British botanists who supported the rule and continental and neo-American botanists who opposed the same.

**Rochester Code (1892)**

A set of rules for plant nomenclature prepared by N. L. Britton and a group of botanists was discussed at the meeting of the American Association for the advancement of Science held in Rochester, New York in 1892. In this Code it was established that each species must be based on a specified herbarium specimen to be known as the *nomenclatural type*. It advocated strict application of the rule of priority and had no legislation comparable to the Kew-rule. The Code accepted all binomials, even tautonyms (the specific epithet repeating the generic name), caused by the retention of the earliest specific epithet when the generic name of a plant was changed. Page priority or the precedence of a name within a publication was also taken into consideration in the application of the rulc of priority.

**Vienna Code (1906)**

The third International Botanical Congress, which met in Vienna in June 1905 gave extensive consideration to nomenclatural matters. The Paris Code was taken as the basis. The major changes included in the Vienna Code are : Linnaeus' *Species Plantarum* (1753) be considered as the starting point of nomenclature of Vascular Plants, and the descriptions of genera as given by Linnaeus in his 5th edition of *Genera Plantarum* (1754) were accepted to validate the publication of generic names appeared in *Species Plantarum* (1753). A list of generic names to be conserved was included. The requirements of Latin diagnoses for the names of new groups, inadmissibility of tautonyms, and the determination of priority of a name only by the date of publication were adopted in the rules.

**American Code (1907)**

This consisted most of the proposal of Rochester Code and agreed in some respects with Vienna Code. It did not accept the *nomtna generica conservanda* or the rcquirements of Latin diagnoses for new taxa. Type concept was accepted. The supporters of American Code pressurised the committee on nomenclature of the Botanical Society of America to prepare a new "type-basis Code" in 1918. This code differed from the American Code only in minor points.

**Brussels Code (1912)**

The fourth International Botanical Congress which met in Brussels in 1910, established different starting points for priority of names of non-vascular plants. The value of type concept was also recognised.

**Cambridge Code (1935)**

When the fifth International Botanical Congress met at Cambridge in 1930, every attempt was made by T.A. Sprague and M.L. Green to harmonize the basic differences between the Vienna Code and the American Code. The Code adopted the type concept as "the application of names of taxonomic groups is determined by means of nomenclatural types". This was the first Code of nomenclature came into use by the botanists all over the world.

**Amsterdam Code (1947)**

At the sixth International Congress held at Amsterdam in 1935 some of the rules governing the Code were again discussed. It was decided that all diagnoses of plants new to science must be in Latin if published after 1.1.1935 instead of 1.1.1932 as mentoined in Cambridge Code. The Rules of Nomenclature were divided into Chapters and Sections, and these were subdivided into Aticles and Recommendations. A list of *Nomina Generica Conservanda* was included in the Code.

A supplement to this Code was compiled by T.A. Sprague in 1950.

**Stockholm Code (1952)**

There were again lengthy discussions on the Rules, when the seventh International Botanical Congress met in Stockholm in 1950; minor changes and few additions were legislated. The word *taxon* (pl. *taxa*) was introduced for the first time to designate any taxonomic group or entity. It was further added that taxon of any rank below the rank of a species which includes the type of the species is to be designated by repeating the specific epithet.

**Paris Code (1956)**

The eighth International Botanical Congress met in Paris in July 1954. The general procedure followed in this Code was as that followed for the Stockholm Code (1952). The nomenclature section decided that the Code should be published in English, French and German languages. One of the main new features of this Code was the separation of the Preamble and Principles from the Rules and Recommendations. Appendix III (*Nomina generica conservanda et rejicienda*) was amended and supplemented. Appendix IV (Determination of types) was reprinted verbatim in accordance with the definite decision to do so taken by the Nomenclature Section at Paris.

**Montreal Code (1961)**

The nineth International Botanical Congress was held in August 1959 at Montreal. In general, the procedure adopted in the preparation of Montreal Code was the same as that followed for Stockholm and Paris. The proposals concerning the question of possible *Nomina specifica conservanda* did not result in any legislative action but were followed by an attempt to assess first the real scope of the problem before changing the rules. The main body of the Code was not amended to any considerable degree. In order to facilitate the use of the Code, the committee decided to retain as far as possible the numbers of the Articles and Recommendations of the Paris Code. Some of the examples in the previous edition of the Code were deleted because they were found to be unsuitable or inaccurate. Names of hybrids and some special categories were amended slightly. The former Appendix II (Special provisions concerning fossil plants) found in Paris Code was not included in that Code but it was decided to incorporate these 'special provisions' where necessary in the main body of the Code. It was made clear that the naming of fossil plants should follow the same lines as that of recent ones. In the Montreal Code *Nomina familiarum conservanda* for Angiospermae as Appendix II, was introduced.

**Edinburgh Code (1966)**

The tenth International Botanical Congress met in August 1964 at Edinburgh. The procedure that followed for Stockholm, Paris and Montreal Codes was adopted in this Code. No numbers of Articles were changed from the Paris and Montreal Codes and only a few changes were effected in the numbering of Recommendations, mainly due to some deletions. The nomenclature of bi-and multi-generic hybrids was amended. *Nomina familiarum conservanda* was slightly amended and the names of families of Recent Gymnosperms were also added. The Edinburgh Congress set up a committee to draw up a glossary of the main technical terms occurring in the Code and that resulted in "*An annotated glossary of botanical nomenclature*" (McVaugh *et al.* 1968).

## Seattle Code (1972)

The eleventh International Botanical Congress met in Seattle in August 1969. Most of the proposals submitted to the Congress were concerned with refinements and increased precision rather than with sweeping changes. Some amendments in the list of *Nomina conservanda* have been given in the Code. Special committees were appointed to slove some questions on the status of names of taxa above the rank of family and the applications of Articles 69, 70 and 71 and to report to the Leningrad Congress of 1975. The main issues at Seattle were the tautonymous designations of taxa between genus and species and below species, the perennial question of superfluous names, and a major reorganization of the rules for hybrids. Articles 22 and 26 have been refined and the word autonym (automatically established names) has been introduced in this Code.

## Leningrad Code (1978)

The Leningrad Code, prepared and edited under the Chairmanship of F.A. Stafleu, was based upon the decisions reached by the Nomenclature Section of 12th International Botanical Congress held in Leningrad in July 1975. This Code which is the current one shows only small differences from the Seattle Code. The numbering of Articles and Recommendation is that of the previous Code. The changes made in this Code are:

(a) Individual paragraphs of all Articles and Recommendations are numbered in a decimal-like system; some paragraphs have been rearranged.

(b) The concept of organ—genera is eliminated for fossil plants (Art. 3).

(c) The principle of automatic typification is extended to those names of taxa above the rank of family that are ultimately based on generic names; application of the principle of priority is recommended when selecting among names thus typified (Art. 16, Rec. 16A and 16B).

(d) The Code makes clear that it does not apply to names of organisms treated as bacteria, but that it does apply to all other organisms treated as plants (Preamble 7, Principle I, Art. 45.4 and Art. 65).

(e) A new name or combination published before 1953 without indication of rank is considered validly published but inoperative in questions of priority except for homonymy and for certain names to be accepted at the rank of variety (Art. 35).

(f) Art. 69 of the former Code is modified on the basis of type method. Now it reads as "A name must be rejected if it has been widely and persistently used for a taxon not including its type. Names thus rejected shall be placed on a list of *Nomina rejicienda*. (Art. 69.1). Also Articles 70 and 71 of the previous Codes dealing with discordant elements and monstrosities were deleted but the numbers of Articles are retained in order to facilitate the use of the Code.

(g) A thorough rewriting of the section on Orthography is presented (Art. 73, Rec. 73C, Rec. 73G and Art. 75).

3

# POLYNOMIAL AND BINOMIAL SYSTEMS

**Polynomial System :**

This system was used in the naming of plants until the first half of 18th century. Polynomial means a plant name consisting of several words in a series, revealing the diagnostic characters of the plant. For example, *Sida cordifolia* L. (1753) was named as '*Althaea maderaspathana, subrotundo folio molli & hirsuto multipilis*' by Plukenet (1692). This system proved difficult for remembering and indexing plant names.

**Binomial System :**

Polynomial system was superseded by the binomial system. Binomial means a name of species in a combination of two words : the first denotes the name of a genus and second the specific epithet. These two words together form a binomial. The generic name is written with a capital initial letter and specific name with small initial letter.

The first application of binomial system to the Vegetable Kingdom as a whole was done by Linnaeus in his *Species Plantarum* (1753) wherein each species is given the usual diagnostic phrase together with the trivial name (specific epithet) written in the the margin of the text against each species. He used this trivial name as a single catchword easy to remember and for indexing thereby avoiding confusion in plant names.

According to Article 23, the name of a species is a binary combination consisting of the name of the genus followed by a single specific epithet. If an epithet consists of two or more words, these must be united or hyphened :

*e.g.* *Tylophora subramanii*

*Strychnos nux-vomica*

According to Article 24, the use of a binary combination for an infraspecific taxon is not admissible. The name of an infraspecific taxon (*viz.* subspecies, variety, etc.) is a combination of the name of a species and an infraspecific epithet connected by the term denoting its rank.

*e.g.* *Crotalaria willdenowiana* subsp. *glabrifoliolata*

*Neanotis monosperma* var. *tirunelvelica*

4

## BROAD OUTLINE OF LENINGRAD CODE (1978)

Preamble

Division I Principles (1-6).

Division II Rules and Recommendations (Arts. 1-75) :

- Chapter I Ranks of taxa, and the terms denoting them (Arts. 1-5).
- Chapter II Names of taxa (general provisions) (Arts. 6-15).
- Chapter III Nomenclature of taxa according to their ranks (Arts. 16-28).
- Chapter IV Effective and valid publication (Arts. 29-50)
- Chapter V Retention, choice, and rejection of names and epithets (Arts. 51-72).
- Chapter VI Orthography of names and epithets and gender of generic names (Arts. 73-75).

Division III Provisions for modification of the Code.

Appendix I Names of hybrids.

Appendix II Nomina familiarum conservanda.

Appendix III Nomina generica conservanda et rejicienda.
Guide for the determination of types.

## MAIN DIVISIONS OF THE INTERNATIONAL CODE OF BOTANICAL NOMENCLATURE (1978)

The Code is divided into three major divisions :

(1) Principles (2) Rules and Recommendations (3) Provisions for modification of the Code. In addition to these there are 3 appendices and a last chapter on "A guide for the determination of types."

The *Principles* form the basis of the system of botanical nomenclature. There are six principles as follows :

1. Botanical nomenclature is independent of zoological nomenclature. The Code applies equally to names of taxonomic groups treated as plants whether or not these groups were originally so treated. (For the purpose of this Code, the word 'plants' does not include the bacteria).

2. The application of names of taxonomic groups is determined by means of nomenclatural types.

3. The nomenclature of a taxonomic group is based upon priority of publication.

4. Each taxonomic group with a particular circumscription, position, and rank can bear only one correct name, the earliest that is in accordance with the Rules, except in specified cases.

5. Scientific names of taxonomic groups are treated as Latin regardless of their derivation.

6. The rules of nomenclature are retroactive unless expressly limited.

The *Principles* do not give detailed rules about nomenclature but show the main ideas that have guided the compilers of the Code.

The *Rules or Articles* give detailed prescriptions to put the nomenclature of the past into order and to provide for that of future; names contrary to a rule cannot be maintained. There are 75 Articles dealt in the Code. These are formed for providing a stable method of naming taxonomic groups, avoiding and rejecting the use of names which may cause error or ambiguity or throw science into confusion.

The *Recommendations* deal with subsidiary points. Their object is to bring about greater uniformity and clearness, especially in future nomenclature. Recommendations are often practical application of the rules. They do not effect the validity of a name but it would be unwise to neglect them.

The Rules and Recommendations apply to all organisms treated as plants (including fungi), whether recent or fossil, with the exception of the bacteria.

The provisions regulating the modification of the Code form its last division.

Provisions for the names of hybrids appear in Appendix I of the Code ; however, special provisions for naming infrageneric and infraspecific ranks for plants used in agriculiure, forestry, and horticulture (and arising either in nature or in cultivation) are provided in the *International Code of Nomenclature of Cultivated Plants—1969*.

6

# RANKS AND NOMENCLATURE OF TAXA

Taxonomic groups of any rank will be referred to as *taxa* (singular: *taxon*) in the Code. The term *taxon* was first presented in its present meaning at the Utrecht symposium on nomenclature during 1948, and first appeared as a part of the Stockholm Code (1952) as decided at the International Botanical Congress held in Stockholm in 1950. The term '*taxon*' is not a category in the classification system but it is a very useful general term for a category on taxonomic group of any rank within the system of taxonomic categories.

The rank of species is basic; one or more species make up a genus; one or more genera make up a family (*familia*) ; one or more families are included in an order (*ordo*) ; one or more orders make up a class (*classis*) ; one or several classes make up a division (*divisio*) ; several divisions make up the kingdom (*regnum*). The ranks of taxa are also further divided by adding the prefix *sub* to the terms denoting the ranks or by the introduction of supplementary terms. A plant may, be assigned to taxa of the following subordinate ranks : *Regnum Subregnum, Divisio, Subdivisio, Classis, Subclassis, Ordo, Subordo, Familia, Subfamilia, Tribus, Subtribus, Genus, Subgenus, Sectio, Subsectio, Series, Subseries, Species, Subspecies, Varietas, Subvarietas, Forma, Subforma* (Art. 4.1).

Names of taxa above the rank of family are automatically typified if they are ultimately based on generic names. Hence the name of the order based on the family Magnoliaceae should end in '*ales*' *i.e.* Magnoliales; and the name of the class based on the order, Magnoliales should end in '*opsida*' *i.e.* Magnoliopsida; and the name of the division based on the class, Magnoliopsida should end in '*phyta*' *i.e.* Magnoliophyta.

The name of a family is a plural adjective used as a substantive; and the name of the family based on the stem of a genus should end in '*aceae*' *i.e.* Magnoliaceae (from *Magnolia*). There are some exceptions with irregular terminations for 8 families by long usage accepted as validly published. However the rule (Art. 18.5) authorizes the substitution of alternative names ending in '*aceae*'. The names of these families with alternative names are as follows: Palmaee (Arecaceae; type, *Areca* L.); Gramineae (Poaceae; type, *Poa* L.); Cruciferae (Brassicaceae; type, *Brassica* L.); Leguminosae (Fabaceae; type, *Faba* Mill.); Guttiferae (Clusiaceae; type, *Clusia* L.): Umbelliferae (Apiaceae; type, *Apium* L.); Labiatae (Lamiacae; type, *Lamium* L.); and Compositae (Asteraceae; type, *Aster* L.). When the Papilionacea (Fabaceae; type, *Faba* Mill.) are regarded as a family distinct from the remainder of the Leguminosae, the name Papilionaceae is conserved against Leguminosae. Accordingly it is wrong to say, for instance, Magnoliaceae is a primitive family; the correct way is to say 'Magnoliaceae are a primitive family'. Any category above the rank of genus should be treated as plural.

The name of a subfamily is formed by adding the termination '*oideae*' to the stem of a genus—*e.g.* Amaranthoideae (from *Amaranthus*) Rosoideae (from *Rosa*); tribes with the termination '*eae*' *e.g.* Amarantheae (from *Amaranthus*), Roseae (from *Rosa*); and subtribe with the termination '*inae*'—*e.g.* Amaranthinae (from *Amaranthus*). If a name of any of the above categories has been published with an improper Latin termination, the termination must be changed to accord with the rule, without change of the author's name. However, when the rank of the group is changed by a later author, he is then cited as author for the name with the appropriate termination.

The name of a genus is a substantive in the singular number, or a word treated as such. It may be taken from any source whatever, and may even be composed in an absolutely arbitrary manner (Art. 20.1) *e.g.* *Impatiens*, *Rosa*. The generic names may not coincide with a technical term currently used in morphology unless it was published before 1.1.1912 and was accompanied by specific epithet. Names such as *Radix*, *Folium* etc. can not now be validly published as new generic names. The generic name may not consist of two words, unless these words are joined by a hyphen.

The names of intergeneric hybrids are formed by combining the names of the two parent genera connected by the multiplication sign ×. e.g. × *Agropogon* (=*Agrostis* × *Polypogon*).

The name of a subdivision of a genus is a combination of a generic name and a subdivisional epithet connected by a term (subgenus, section, series, etc.) denoting its rank (Art. 21.1). The epithet is either of the same form as a generic name or a plural adjective agreeing in gender with the generic name and written with a capital initial letter. The epithet should not end with—'*oides* or '*opsis*', or begin with '*Eu*' *e.g. Costus* subg. *Metacostus*; *Euphorbia* sect. *Tithymalus*; *Euphorbia* subsect. *Tenellae*.

The name of a species is a binary combination consisting of the name of the genus followed by a single specific epithet. If an epithet consists of two or more words, these are to be united or hyphened. The epithet of a species may be taken from any source whatever, and may even be composed arbitrarily. *e.g. Zenkeria sebastinei*; *Strychnos nux-vomica*.

The name of an infraspecific taxon is a combination of the name of a species and an infraspecific epithet connected by a term denoting its rank. The use of a binary combination for an infraspecific taxon is not admissible. *e.g. Crotalaria willdenowiana* subsp. *glabrifoliolata*; *Leucas lavanduliifolia* var. *nagalapuramiana*.

# 7

## DEFINITIONS

**Valid Publication of a Name :**

A name published in accordance with Arts. 32-45 of ICBN (*vide* Chapter 11). A name of a taxon has no status under the "Code" unless it is validly published. A validly published name only can be treated as legitimate or illegitimate.

The term "invalid" though does not appear in the Code, usually refers to a name which is not validly published. Thls term however is sometimes used erroneously as synonymous with illegitimate.

**Legitimate Name :**

A validly published name or epithet that is in accordance with all the rules.

**Illegitimate Name :**

A validly published name or epithet that is contrary to one or more Articles of the Code. An illegitimate name must be rejected under the Code. Some of the common illegitimate names are :

Superfluous names

Later homonyms

Tautonyms

Rejected names (*nomina rejicienda*) : A name must be rejected if it has been widely and persistently used for a taxon not including its type. *i.e.* a name that is used in different senses and has become a long persistent source of error or confusion.

**Correct Name :**

A legitimate name that must be adopted under the rules for a taxon with a particular circumscription, position and rank. A legitimate name may be correct or incorrect according to different concepts of the taxa.

For example, the generic name *Hibiscus* L. (Sp. Pl. 693. 1753; Gen. Pl. ed. 5. 310. 1754) is legitimate as it is in accordance with the rules. The same is true of the generic name *Abelmoschus* Medicus (Malv. 46. 1787). Both generic names are correct when the genera are kept separate. Augustin de Candolle (Prodr. 1 : 449. 1824) united *Abelmoschus* Medicus and *Hibiscus* L. in a single genus and when this treatment was accepted, as followed in most of the earlier Indian floras, *Hibiscus* L. was the only correct name with this particular circumscription. Subsequently, however, K. Schumann (in Engler & Prantl, Nat. Pflanzenfam. 3 (6) : 47. 1895) and following him several recent monographers re-established the genus *Abelmoschus* Medicus. The legitimate generic name *Abelmoschus* Medicus may therefore be correct or incorrect according to different concepts of the taxa.

**Synonym :**

Synonyms are the different names used for the same taxonomic group or taxon. A synonym, as commonly used is meant for a later synonym or a synonymous name that is to be rejected. There are two kinds of synonyms : Nomenclatural synonyms and Taxonomic synonyms.

*Nomenclatural synonyms* are the names based on the same type ; often called as homotypic synonyms.

For example, *Miliusa tomentosa* (Roxb.) Sinclair, *Saccopetalum tomentosum* (Roxb.) Hook. f. & Thoms. and *Uvaria tomentosa* Roxb.

are names based on the same type and hence they are nomenclatural synonyms. There is no question on their identity.

*Taxonomic synonyms* are the names based on different types considered to belong to the same taxon; often called as heteratypic synonyms.

For example, *Hibiscus vitifolius* L. (1753) and *Hibiscus obtusifolius* Willd. (1801) are names based on two different types [Linnaeus n. 265, and Klein *s.n.* (B-W n. 12899) respectively]. An examination of both the type materials has proved them identical. This, of course, is entirely based on a taxonomic judgement. Hence *Hibiscus obtusifolius* Willd. is a taxonomic synonym of *Hibiscus vitifolius* L.

**Homonym :**

A name spelled exactly like a validly published name for a taxon of the same rank based on a different type.

*e.g.* *Jatropha heterophylla* Steudel, Nom. Bot. ed 2. 1: 799. 1840, and *Jatropha heterophylla* Heyne ex Hook. f. Fl. Brit. India 5: 382. 1887.

The later homonym *i.e. J. heterophylla* Heyne ex Hook. f. is illegitimate and is replaced by the new name—*J. heynei* Bal. (in Bull. bot. Surv. India 3: 41. 1961).

**Tautonym :**

A binary name in which the specific epithet exactly repeats the generic name with or without transcribed symbol.

*e.g. Malus malus*

Tautonyms are treated as illegitimate in botanical nomenclature. It may be noted that names such as *Cajanus cajan* and *Bambusa bambos* are not considered as tautonyms as the specific epithets do not repeat exactly the generic names.

**Autonym :**

A name automatically established for a taxon, merely as a consequence of the proposal of another taxon. For example, the recognition of a subspecies or variety that does not contain the type of the species automatically establishes a subspecies or variety that does contain the type. Hence the final epithet of an autonym at the infraspecific level is that of the specific name unaltered, but without citation of an author's name.

*e.g. Abutilon indicum* (L.) Sweet subsp. *indicum.*

The publication in 1965 of the name *Crotalaria willdenowiana* subsp. *glabrifoliolata* Ellis, automatically established the name of another subspecies *Crotalaria willdenowiana* DC. subsp. *willdenowiana*, the type of which is that of the name *Crotalaria willdenowiana* DC.

**Comb. Nov. (Combinatio Nova) :**

This means new combination. It is a new name formed from a previously published name or epithet and is typified by the type of the basionym. When a taxon of lower rank than genus is transferred to another genus or species, with or without alteration of rank, based on taxonomic judgement, the original name or epithet if legitimate (basionym) is retained and the author of the basionym is cited in parentheses followed by the name of the author who effected the alteration (the author of the new combination).

For example, *Miliusa tomentosa* (Roxb.) Sinclair is a new combination based on *Uvaria tomentosa* Roxb.

When the rank of the taxon is altered during the transfer and its epithet from the name in the old rank has been retained, the expression *stat. nov.* (*status novus*) is given for the new combination.

For example, *Abutilon indicum* (L.) Sweet subsp. *albescens* (Miq.) Borrs. is a new subspecific status (*stat. nov.*) given for the species *Abutilon albescens* Miq.

**Nom. Nov. (Nomen Novum) :**

This means new name. A new name is usually a substitute to replace an earlier name and is typified by the type of the earlier name. This is often revealed as such only by the citation of the older name in the synonymy (replaced synonym). A new name is chosen only when no legitimate name is available.

For example, *Baeolepis* Decne. ex Moq. [in DC. Prodr. 13 (2) : 216. 1849] is a new name or an avowed substitute for the later homonym *Brachylepis* Wt. & Arn. (in Wt. Contrib. Bot. Ind. 63. 1834).

*Tabernaemontana gamblei* Subr. & Henry [in Bull. bot. Surv. India 12 : 5. 1970 (1972)] is a new name for *Ervatamia caudata* Gamble; *Ervatamia caudata* Gamble (in Kew Bull. 1921 : 310. 1921) when transferred to the genus *Tabernaemontana* must be called *Tabernaemontana gamblei* Subr. & Henry (1972) because of the existence of the name *Tabernaemontana caudata* Merr. (1909), referring to a different species. In other words, the epithet *caudata* is preoccupied in the genus *Tabernaemontana.*

*Emblica officinalis* Gaertn. (Fruct. 2 : 122. 1791) is a new name for *Phyllanthus emblica* L. *Phyllanthus emblica* L. Sp. Pl. 982. 1753 when transferred to the genus *Emblica*, must be called *Emblica officinalis* Gaertn. (1791), because Gaertner could not retain the specific epithet "emblica" as the resulting binary name *Emblica emblica* is a tautonym which is inadmissible in accordance with the Code. Gaertner, therefore, called the plant by the new name (*nom. nov.*)—*Emblica officinalis* Gaertn. It may be noted that in describing and naming a taxon for the first time, an author usually gives a name, not a new name.

**Latin Diagnosis :**

Diagnosis is a brief statement of characters which differentiates the taxon from its nearest relatives. The diagnosis of a species or an infraspecific taxon may be very short, merely a statement of one or two characters not possessed by the other species in the same genus. According to the Code a name of a new taxon (the algae and fossils excepted) published on or after 1 January 1935 must be accompanied by a Latin description or a diagnosis or by a reference to a previously and effectively published Latin description or diagnosis of the taxon. It is recommended that authors publishing names of new taxa of recent plants should give or cite a full description in Latin in addition to the diagnosis.

# 8

## THE TYPE METHOD OR TYPIFICATION

The Code lays great emphasis on typification of various taxa in order to achieve stabilization of names. A concept of types in the application of names was started in America; the Rochester Code (1892) established the concept of types as a scientific and fundamental basis to determine the precise application of names. The American Code (1907) was predicated on the acceptance of the type concept. Further in 1918, the Committee on Nomenclature of ihe Botanical Society of America formulated a Type-basis Code largely based on this concept.

Though the International Rules (Vienna Code, 1905) as formulated at Vienna ignored the concept of types, the fourth International Congress which met at Brussels in 1910, gave recognition of the value of the type concept. In Cambridge Code (1930) it was included as an Article as "Art. 18. The application of names of taxonomic groups is determined by means of nomenclatural types..." In Recommendation IV under the same Article, it is said: "When publishing names of new groups, authors should indicate carefully the subdivision which is the type of thc new name : the type genus in a family, the type species in a genus, the type variety or specimen in a species..." The Stockholm Code (1952) introduced a certain number of definitions on types (holotype, isotype, paratype, etc.). A great emphasis was laid on types among the fundamental principles of the Paris Code (1956). Principle II states : "The application of names of taxnomic groups is determined by means of nomenclatural types". Article 35, of the Paris Code, for the first time, insisted on the indication of the type for validity : "Publication on or after 1 January,

1958 of the name of a new taxon of recent plants of the rank of order or below is valid only when the nomenclatural type is indicated." Further, it was accepted as a recommendation that when the nomenclatural type of a new taxon is a specimen, the place where it is permanently preserved must be indicated. According to Article 7 of Montreal Code (1961), the application of names of taxa of the rank of family or below is determined by means of nomenclatural types. This has been continued up till Leningrad Code (1978). Recently, in Leningrad Code (1978), it has been further added that the application of names of taxa above the rank of a family is also determined by the types when the names are ultimately based on generic names. In other words the principle of autómatic typication is extended to those names of taxa above the rank of family that are ultimately based on generic names. Typification is dealt in Leningrad Code (1978) under Arts. 7—10.

A nomenclatural type is that element to which the name of a taxon is permanently attached, wheather as a correct name or as a synonym. The nomenclatural type is not necessarily the most typical or representative element of a taxon. The type of a genus or any taxon between genus and species is a species, that of a family or of any taxon between family and genus is the genus. The principle of typification does not apply to names of taxa above the rank of family, except for names that are automatically typified by being ultimately based on generic names. Type of a name of taxon assigned to a group with a nomenclatural starting-point later than 1753 is to be determined in accordance with the indication or description and other matter accompanying its first valid publication.

The various kinds of types acceptable to the Code are treated below :

**Holotype** : A holotype is the one specimen or other element used by the author designated by him as the nomenclatural type (Art. 7.3). A nomenclatural type is that element to which the name of a taxon is permanently attached, whether as a correct name or synonym. As long as holotype is extant, it automatically fixes the application of the name concerned. The type (holotype, lectotype or neotype) of a name of a species or infraspecific taxon is a single specimen or other element except in the following case : for small herbaceous plants and for most non-vascular plants, the type may consist of more than one individual, which ought to be conserved permanently on one herbarium sheet or in one preparation. The holotype is chosen by the original author from a single gathering made by a collector at one time, and expressed definitely at the time of the original publication. Publication on or after 1.1.1958 of the name of a new taxon of the rank of family or below is valid only when the nomenclatural type is indicated.

*e.g. Polygala jacobii* Chandr. in Bull. bot. Surv. India 9 (1-4): 288-289. 1967.

Holotypus *Chandrabose* 28777 A (MH) *vide* Plate-1.

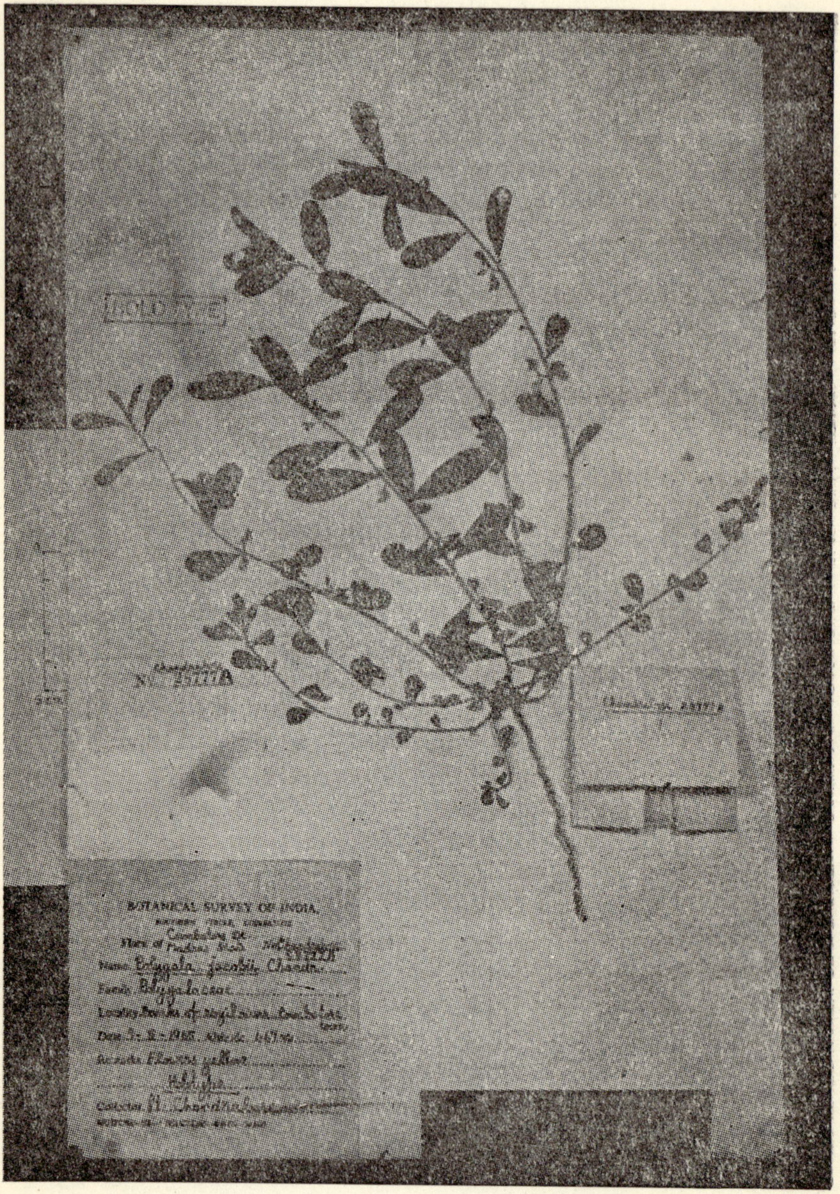


PLATE—1 : *Polygala jacobii* Chandr. Holotypus *Chandrabose* 28777A (MH)

**Isotype** : An isotype is any duplicate (part of a single gathering made by a collector at a time) of the holotype (Art. 7.6). For instance, if a new species is based on a single collection (consisting of 5 specimens and 5 herbarium sheets made out of them) bearing the same field number, the original author will designate one among them as holotype ; and the rest of the 4 specimens will be the isotypes.

*e.g.* *Polygala jacobii* Chandr. in Bull. bot. Surv. India 9 (1-4) : 288-289. 1967.

Isotype *Chandrabose* 28777 B-K (MH).

**Paratype** : A paratype is a specimen cited in the protologue other than the holotype, isotype(s) or syntypes. If a new taxon is described based on more than one gathering made at different times, the author will designate one gathering as holotype and isotype(s). The rest of the gathering bearing different field numbers form the paratypes. For instance, if the new taxon is collected with flowers in one season, and in fruit in other season, these two gatherings will bear two different field numbers. While describing the new taxon, the author(s) will choose one gathering for holotype and istotypes; the next gathering cited in the protologue forms the paratype.

*e.g.* *Polygala jacobii* Chandr. in Bull. bot. Surv. India 9 (1-4) : 288-289. 1967.

Paratypus *Rangachari s. n.* (Acc. no. 2444, MH) *vide* Plate-2.

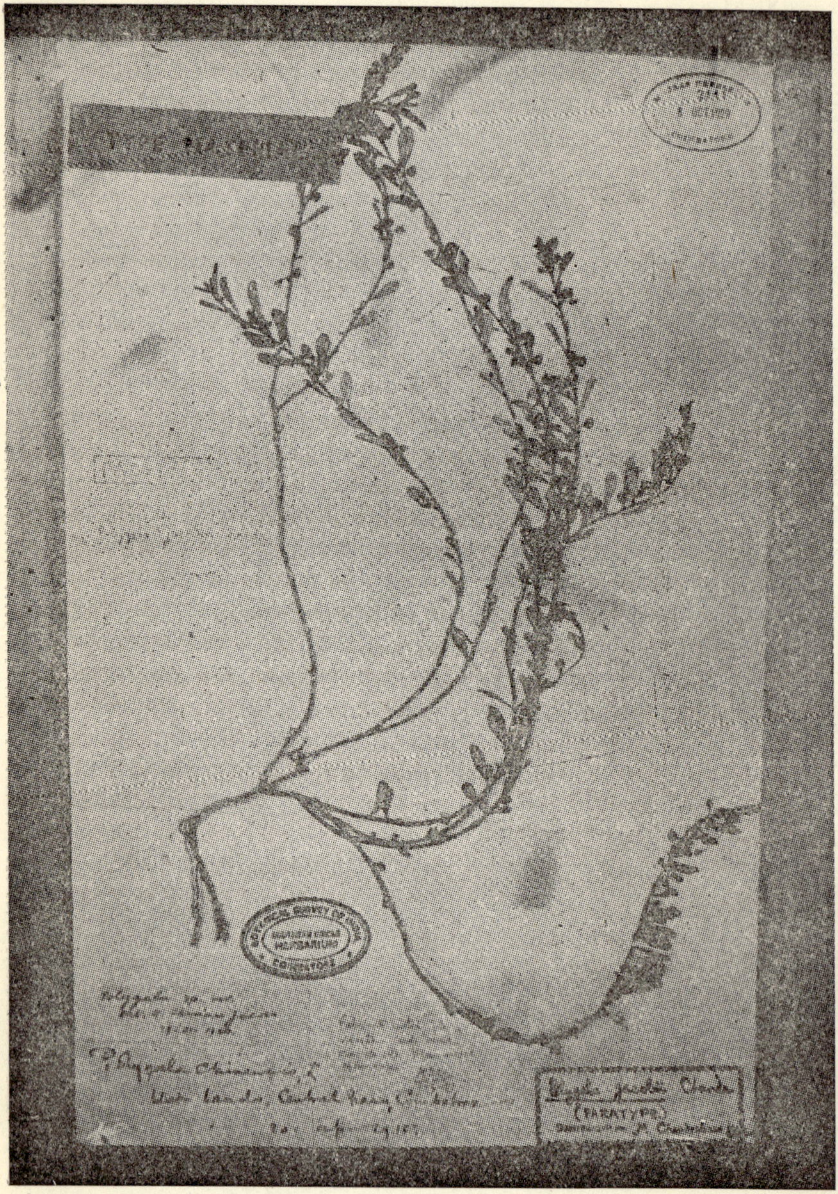

PLATE--2 : *Polygala jacobii* Chandr. Paratypus *Rangachari* s. n. (Acc. No. 2444. MH)

**Syntype** : A syntype is any one of two or more specimens cited by the author when no holotype was designated, or any one or two or more specimens simultaneously designated as types. (Art. 7.7)

**Lectotype** : A lectotype is a specimen or other element selected from the original material to serve as a nomenclatural type when no holotype was designated at the time of publication or as long as it is missing. When two or more specimens have been designated as types by the author of a specific or infraspecific name, the lectotype must be chosen from among them (Art. 7.5). Designation of lectotype should be undertaken only in the light of an understanding of the group concerned. In choosing a lectotype, all aspects of the protologue should be considered as a basic guide. If a holotype is lost or destroyed, the lectotype is chosen from the isotype(s). If no holotype was designated and if syntypes exist, one of them must be chosen as lectotype. If neither an isotype, nor a syntype is extant, a paratype if such exists, may be chosen as lectotype.

*e.g.* 1. In 1826, Blume published the new genus *Mastixia* (Bijdr. 654. 1826) with the species *M. pentandra* Bl. and *M. trichotoma* Bl. The type species of the genus, however, was not indicated at the time of original publication. Hutchinson (Gen. Fl. Pl. 2 : 45. 1967) chose *M. pentandra* Bl. as the lectotype of the genus and his choice should be retained.

*e.g.* 2. Rangachari and Tadulingam (*J. Bombay nat. Hist. Soc.* 24: 846. 1916) while describing the new species *Cynodon barberi* cited several specimens in the protologue, but failed to indicate one as the type or holotype. Hence all the specimens cited became syntypes. Jain *Bull. bot. Surv. India* 8 : 204. 1967) after a critical study, correctly selected one of the specimens—"Madras Herbarium, South India Flora No. 13715", deposited in CAL, as the lectotype.

**Neotype** : A neotype is a specimen or other element selected to serve as nomenclatural type as long as all of the material on which the name of the taxon was based is missing (Art. 7.8). A neotype may be designated only when all the originally cited material is believed lost or destroyed. In selectiag a neotype more care and critical knowledge are essential as the reviewer usually has no guide except his own judgement as to what best fits the protologue.

*e.g.* The new species *Farsetia macrantha* was described by Blatter and Hallberg ( *J. Bombay Nat. Hist. Soc.* 26 : 220. 1918); in the Protologue they cited 3 numbers *viz.* Blatter 5785, 7300 & 7305. These specimens presumably were deposited in the Blatter Herbarium, Bombay (BLAT), where the entire collection of Blatter, including the 'types' have been preserved. Bhandari [*J.*] *Bombay Nat. Hist. Soc.* 72. 606, 1975 (1976)] while working on this species could not trace the type specimens in the Blatter Herbarium, and concluding that all the original material has, presumably been, lost or destroyed", designated *Bhandari* 507, as neotype, "exemplifying the true application of the name *F. macrantha* Blatt. & Hallb. and to ascertain with certainty the sense in which the same must be used. The specimen comes from the type locality and is in complete agreement with the original diagnosis."

9

## GUIDE FOR THE DETERMINATION OF TYPES

The determination or selection of the nomenclatural types of previously published taxa needs certain guidelines as follows :

A holotype designated by the original author in the original publication of the name of the taxon is final and should not be changed. If he had indicated only one element, that must be accepted as the Holotype. It is to be noted that neither the original author of a taxon, nor any one else can choose a holotype at any time after the publication of the taxon; any such later choice always establishes a lectotype.

A nomenclatural type for a new name or epithet published as an avowed substitute (nomen novum) is the same type of the older name.

A lectotype is selected from the original material on which a taxon was based; in other words, a lectotype must be chosen from among elements that were definitely studied by the author up to the time the name of the taxon was published and included in the protologue. If a holotype was designated by the original author and has been lost or destroyed, an isotype, if such exists, must be chosen as the lectotype. If no holotype was designated by the original author and if syntypes exist, one of them must be chosen as the lectotype. If neither an isotype nor a syntype is extant, a paratype if such exists, may be chosen as the lectotype. The first choice of a lectotype must be followed by subsequent workers unless the original material is rediscovered, or unless the choice made was proved incorrect.

A neotype is designated only when all the originally cited materials are believed lost or destroyed. In selecting a neotype more care and critical knowledge are essential, as the reviewer usually has no guide except his own judgement as to what best fits the protologue. A lectotype always takes precedence over a neotype.

# 10

## PRIORITY

The six principles (Division I of the Leningrad Code, 1978) form the basis of the System of Botanical Nomenclature. Among them, Principle III and IV deal with the principle of priority in particular, and correspondingly Arts. 11 to 14 give detailed prescriptions on priority and limitation of the principle of priority. Each family or taxon of lower rank with a particular circumscription, position, and rank can bear only one correct name, special exceptions being made for 8 families for which alternative names are permitted (*vide* Chapter 6).

For any taxon from family to genus inclusive, the correct name is the earliest legitimate one with the same rank except in cases of limitation of priority by conservation. For any taxon below the rank of genus, the correct name is the combination of the earliest available legitimate epithet in the same rank with the correct name of the genus or species to which it is assigned. The principle of priority is not mandatory for names of taxa above the rank of family.

The principle of priority was originally intended to give stability for plant names, but its strict application causes serious instability in plant nomenclature as and when a Botanist unearths some old and forgotten publications. However, there would be no further changes when all the published records are once checked up carefully.

A name of a taxon has no status in the "Code" unless it is validly published. Valid publication of names of plants of different groups is treated as beginning at some prescribed dates corresponding to the publication of some works.

For example, Spermatophta and Pteridophyta, 1 May 1753 (Linnaeus, *Species Plantarum* ed. 1.).

Generic names which first appear in Linnaeus' *Species Plantarum* ed. 1. (1753) and ed. 2 (1762-63) are associated with the first subsequent description given under these names in Linnaeus' *Genera Plantarum* ed. 5 (1754) and ed. 6 (1764).

The two volumes of Linnaeus' *Species Plantarum* ed. 1 (1753), which appeared in May and August respectively, are treated as having been published simultaneously on the former date (1 May 1753). *e.g.* The generic names *Thea* L. Sp. Pl. 515 (May 1753) and *Camellia* L. Sp. Pl. 698 (August 1753) are treated as having been published simultaneously in May 1753. Sweet (Hort. Suburb. Lond. 157. 1818) who was the first to unite the two genera, chose *Camellia* for the combined genus, citing *Thea* as a synonym. The above example indicates that the Code does not recognise priority of position, provided both names were published on the same date.

The Code provides in Appendices II and III lists of names of conserved families and genera respectively. These are exceptions to the strict application of the rules of priority (*vide* Chapter 18).

# EFFECTIVE AND VALID PUBLICATION

**Effective Publication :**

In the present Code (Leningrad Code, 1978), conditions and dates of effective publication of names of taxa have been dealt in the Arts. 29–31.

These are summarised as follows :

Publication should be in printed matter; but is effective by indelible autograph if done before 1.1.1953. The printed matter must be widely distributed to the general public or atleast to botanical institutions with libraries accessible to botanists. Publication on or after 1.1.1953 of a new name in tradesmen's catalogues or non scientific newspapers, and on or after 1.1.1973 in seed exchange lists, does not constitute effective publication.

It is not effected by communication of new names at a public meeting, or by placing of names in collections or gardens open to public, or by the issue of microfilm made from manuscripts, typescripts or other unpublished material.

The date of effective publication is the date on which the printed matter became available. The distribution on or after 1.1.1953 of printed matter along with a herbarium specimen does not constitute effective publication.

**Valid Publication :**

The name of a taxon when published must fulfil a number of conditions for valid publication. The conditions and date of valid publication are dealt in Arts. 32—45. The following conditions are to be followed/fulfilled for valid publication :

The name of a taxon must be effectively published. *i.e.* all the conditions of effective publication are to be fulfilled for a valid publication.

The name of a taxon must be accompanied by a description or diagnosis of the taxon or by a reference to a previously and effectively published description or diagnosis of it. Publication of a name should not be validated solely by a reference to a description or diagnosis published before 1753.

A new combination published on or after 1.1.1953, for a previously and validly published name, basionym should be clearly indicated and a full and direct reference given to its author and original publication with page or plate reference and date. A proposed new name (*nomen novum*) published on or after 1.1.1953 for a previously and validly published name is valid only when the replaced synonym is clearly indicated and a full and direct reference given to its author and original publication with page or plate reference and date.

The proposal of a provisional name in anticipation of future acceptance of the group concerned is not valid.

A clear indication of the rank of the taxon concerned must be given with a new name or combination on or after 1.1.1953.

A name of a new taxon of plants (except algae and fossils) published on or after 1.1.1935 must be accompanied by Latin description or diagnosis or by a reference to a previously and effectively published Latin description or diagnosis of the taxon.

Publication of the name of a new taxon of the rank of family or below on or after 1.1.1958 is valid only when the nomenclatural type is indicated. The place where the type specimen is permanently conserved also should be indicated.

For valid publication of a genus, the name of a genus must be accompanied by a description or diagnosis of the genus or by a reference to a previously and effectively published description or diagnosis of the genus in that rank or as a subdivision of a genus. Monotypic new genus based on a new species is validated by the provision of a combined generic and specific description or diagnosis. The generic names published before 1.1.1908 by the provision of an illustration with analysis showing essential characters are valid.

The name of a species or of an infraspecific taxon published before 1.1.1908 is validly published if it is accompanied only by an illustration with analysis showing essential characters.

# 12

## CITATION OF AUTHORS' NAMES AND OF LITERATURE AND GENERAL RECOMMENDATIONS ON CITATION

The general procedures to be adopted for the above are stated in Arts. 46—50.

The name of a taxon should be followed by the name of the author(s) who first validly published the name concerned.

*e.g.* Annonaceae Juss.

*Annona L.*

*Annona squamosa L.*

Authors' names put after the names of plants may be abbreviated, unless they are very short. One-syllable names, in general, are not to be abridged; however if it is long the first consonants only are given (Fr. for Fries). It is best to conform to a well-established custom in abridging the names provided all confusion is avoided. (L. for Linnaeus; DC. for de Candolle; H.B.K. for Humboldt, Bonpland and Kunth). Hook. f. or Hk. f. is correct for J.D. Hooker; Hook. & Hk. will lead to confusion with the abbreviation of his father's name William Jackson Hooker.

**Use of *et* or *et al.* :**

When a name is published jointly by two authors, the names of both are connected by *et* or by an ampersand (&). It is wrong to use *and* between the authors' names since the name of the plant is in Latin and this includes the authors' names also.

*e.g.* *Zenkeria sebastinei* Henry et Chandrab. (or Henry & Chandrab).

When a name is published jointly by more than two authors, the citation should be restricted to that of the first one followed by *et al.* (*i.e.* and others).

*e.g.* *Indotristicha tirunelveliana* Sharma, Karthikeyan & Shetty (Bull. bot. Surv. India 16: 157-161. 1974) should be cited as *I. tirunelveliana* Sharma *et al.*

**Use of *ex* :**

When an author who first validly publishes a name ascribes it to another person (who has earlier proposed, but not validly published the name) the correct author citation is the name of the actual validly publishing author, but the name of the other person followed by the connecting word *ex* may be inserted before the validly publishing author, if desired. For example, Wight and Arnott in their Prodromus (1834) first validly published the name *Hopea wightiana* but ascribed it to Wallich who proposed the same as a new species in his catalogue (1828-1849) by simply listing the names without any description; Wight and Arnott provided description, etc., and validly published the same. Hence the correct citation should be *Hopea wightiana* Wight & Arn. or *Hopea wightiana* Wall. ex Wight & Arn.

**Use of *in* :**

When a name with a description supplied by one author is published in another author's work, the word *in* should be used to connect the names of two authors.

*e.g.* *Osbeckia wynaadensis* C.B. Clarke in Hook. f. Fl. Brit. India 2: 521. 1879.

**Use of *emend.* :**

If an original description of a taxon is incomplete and subsequently corrected, or the circumscription of a taxon is altered retaining the same type, the citation of the original author is retained, but followed by a word *emend.* and the name of the author responsible for the change. This is an important element of revisionary work.

*e.g.* *Phyllanthus* L. emend. Muell.-Arg.

*Manisuris forficulata* Fischer emend. Jain

**Use of** *pro parte* (*p.p.*) :

It means partly or in part. When an alteration of the circumscription of a taxon has been considerable, resulting in its division into two or more taxa, in the citation of the newly formed taxa the misapplied original name is indicated by adding *pro parte* (*p.p.*).

*e.g.* **Gomphia** Schreb. Gen. Pl. ed. 8. 1 : 291. 1789.

*Ochna* L. Sp. Pl. 513. 1753. *p.p. excl. typus.*

**Maba neilgerrensis** Wight, Illust. 2 : t. 148. 1850.

*M. buxifolia* sensu Clarke in Hook. f. Fl. Brit. India 3 : 551. 1882 *p.p.* non Pers. 1807.

In the above citation *p.p.* indicates that a part of the description of *M. buxifolia* as given by Clarke in Hooker's Flora of British India, pertains to *M. neilgerrensis* Wight. *M. buxifolia* sensu Clarke means *M. buxifolia* as treated by Clarke; and *non* Pers. 1807 means *M. buxifolia* as treated by Clarke is not that of the true *M. buxifolia* Pers. 1807.

**Use of** *sensu amplo* (*s. ampl.*) and *sensu stricto* (*s. str.*) :

*Sensu amplo* or *sensu lato* means in the broad sense, and *sensu stricto* means in the narrow sense.

For example, the genus *Eugenia* L. *sensu amplo* includes species of *Syzygium*, *Cleistocalyx*, *Eugenia*—proper etc. In other words the circumscription of the genus is widened to accommodate the other allied genera. When *Syzygium* and *Cleistocalyx* are treated as distinct genera from *Eugenia*—proper, the latter is to be cited as *Eugenia* L. *sensu stricto*.

**Use of** *parentheses* :

When a genus or a taxon of lower rank is altered in rank but retains its name or epithet, the author of the basionym or the epithet bringing synonym must be cited in parentheses followed by the author who effected the alteration. The same holds when a taxon of lower rank than genus is transferred to another genus or species, with or without alteration of rank.

*e. g.* (i) *Sida retusa* L. when reduced to a subspecies of *Sida rhombifolia* L. becomes *S. rhombifolia* subsp. *retusa* (L.) Borss.

(ii) *Meesia serrata* Gaertn., when transferred to the genus *Gomphia* becomes *Gomphia serrata* (Gaertn.) Kanis.

**Use of** *pro syn.* :

*pro syn.* means as synonym.

In the citation of a name published as a synonym the words *pro syn.* should be added.

When an author has published as a synonym a manuscript name of another author, the word *ex* should be used in citations to connect the names of the two authors.

For example, *Hedyotis eualata*, a manuscript name of Beddome published by Gamble as a synonym of *Oldenlandia eualata* Gamble, should be cited thus : *Hedyotis eualata* Bedd. ex Gamble in Kew Bull. 1919: 405. 1919, *pro syn.* However, this will not constitute valid publication.

**Use of** *nomen nudum* (*nom. nud.*) :

*Nomen nudum* means naked names *i.e.* merely a name without description, etc. In citation of such a name, its status should be indicated by adding *nom. nud.*

*e.g. Convolvulus adpressus* Wall. Cat. no. 1424. 1828, *nom. nud.*

Wallich in his Catalogue (1828-1849) published many new names without any description or reference to an earlier published description. Hence these names are not valid, and most of these were subsequently validated.

**Use of** *auct. non* :

A misapplied name is indicated by the words *auct. non* followed by the name of the original author and the bibliographical reference of the misidentification.

*e.g.* **Phyllanthus fraternus** Webster, Contr. Gray Herb. 176 : 53. 1955. *P. niruri* auct. non L. 1753; Hook. f. Fl. Brit. India 5: 298. 1887.

This plant has been treated as *P. niruri* L. in most of our Indian Floras. Webster's work proved that the Indian plant is only *P. fraternus.* The words *auct. non* L. in the citation mean that *P. niruri* as treated by all authors, except Linnaeus, is *P. fraternus* Webster.

**Use of** *nom. cons.*:

If a generic name is conserved (*nomen conservandum*) the words *nom. cons.* should be added to the citation.

*Tectona* L. f. Suppl. 20, 151. 1781 *nom. cons.*

# 13

## RETENTION OF NAMES OR EPITHETS OF TAXA

Retention of names or epithets of taxa under various conditions are discussed in Arts. 51—56.

An alteration of the diagnostic characters or of the circumscription of a taxon does not warrant a change in its name. For example, the genus *Gomphandra* Wall. ex Lindl. was revised by Beccari who made some alterations in the diagnostic characters and circumscription of the genus; however the generic name must be retained since the type of *Gomphandra* remains in the genus; it is cited as *Gomphandra* Wall. ex Lindl. or *Gomphandra* Wall. ex Lindl. emend. Becc.

When a genus is divided into two or more genera, the generic name must be retained for one of them. Similarly when a species is divided into two or more species, the specific epithet must be retained for one of them. This is applicable to infraspecific taxa also.

When a subdivision of a genus is transferred to another genus without change of rank, its epithet must be retained.

When a species is transferred to another genus or placed under another generic name for the same genus without change of rank, the specific epithet, if legitimate, must be retained or, if it has not been retained, must be reinstated unless one of the following obstacles exists :

(a) The resulting binary name is a later homonym or tautonym.

(b) An earlier legitimate specific epithet is available.

# 14

## CHOICE OF NAMES

The stipulations to be followed for choosing names when the taxa of the same rank are united are given in Arts. 57—58; and when the rank of a taxon is changed in Arts. 60—61.

When two or more taxa of the same rank are united, the oldest legitimate name is retained. For example, K. Schumann (in Engler & Prantl, Pflanzenfam. III. 6 : 5. 1890) uniting three genera *Sloanea* L. (1753), *Echinocarpus* Blume (1825), and *Phoenicosperma* Miq. (1865) rightly adopted the oldest of these three generic names, *Sloanea* L. for the resulting genus.

The author who first unites taxa bearing names for epithets of the same date must choose one of them, and his choice must be followed. For example, Linnaeus published *Waltheria indica* and *W. americana* in Species Plantarum 673. 1753 (*i. e.* on the same date—1st May, 1753). Robert Brown (in Tuckey, Narr. Exp. Congo. 484. 1818) was the first person to unite these two species. He adopted the name *W. indica* L. for the combined species, and this name is accordingly to be retained.

When the rank of a genus or infrageneric taxon is changed, the correct name or epithet is the earliest legitimate one available in the new rank. A name or an epithet will have no priority outside its rank. For example, *Cyperus iria* var. *rectangularis* Kukenth. (in Engler, Pfreich. 101 : 152. 1935) when raised to the specific rank is correctly called *Cyperus alulatus* Kern (in Reinwardtia 1 : 463. 1952) and not *Cyperus rectangularis* (Kukenth.) Bennet (in Indian Forester 95:

692. 1969). However, it is recommended that if there is no earlier name or epithet available in the new rank, the original name or epithet may be retained unless it is contrary to the Code.

When a taxon of a rank higher than the genus and not higher than family is changed in rank, the stem of the name is to be retained and only the termination altered (-inae, -eae,- oideae, -aceae) unless it is contrary to the Code : *e.g.* The subtribe *Drypetinae* Pax (1890) of Euphorbiaceae was raised to the rank of tribe by Hurusawa. Then it became *Drypeteae* (Pax) Hurusawa.

# 15

## REJECTION OF NAMES AND EPITHETS

The conditions under which names and epithets are rejected are furnished in Arts. 62—72.

A legitimate name or epithet should not be rejected simply because it is inappropriate or disagreeable, or because another is perferable or better known or because it has lost its meaning. For example, the generic name *Richardia* L. (1753) is not to be changed to *Richardsonia* as was done by Kunth (Mem. Mus. Hist. Nat. 4 : 430. 1818) although the name was originally dedicated to a botanist, Richardson. The name *Scilla peruviana* L. (Sp. Pl. 309. 1753) is not to be rejected because the species does not grow in Peru.

A name is illegitimate and is to be rejected if it was nomenclaturally superfluous when published. For example, the generic name *Elaeodendron* Jacq. [Icon. Pl. Rar. 1 (2) : t. 48. 1782] is illegitimate because it was a superfluous name for *Cassine* L. (Sp. Pl. 268. 1753); the two genera had precisely the same circumscription. Hence the name *Elaeodendron* Jacq. published at a later date is to be rejected.

A name is illegitimate and must be rejected if it is a later *homonym* *i.e.* if it is spelled exactly like a name previously and validly published for a taxon of the same rank based on a different type. For example, the name *Banalia* Moq. (1849) given to a genus of Amaranthaceae, is a later homonym of *Banalia* Rafin. (1840), a name previously and validly published for a genus of Euphorbiaceae; *Banalia* Moq. is therefore rejected, as was done by A.N. Henry and Bela Roy (Bull. bot. Surv. India 10 : 274. 1968 (1969), who renamed it *Indobanalia*.

Mere orthographic variants of the same name are treated as homonyms when they are based on different types.

Illegitimate epithets must not be taken into consideration for purposes of priority except in the rejection of later homonym. A specific epithet is not illegitimate merely because it was originally published under an illegitimate generic name, but it must be taken into consideration for purposes of priority if the epithet and corresponding combination are in other respects in accordance with the rules. The same way is applicable for an infraspecific epithet published under an illegitimate name of a species.

16

# ORTHOGRAPHY OF NAMES AND EPITHETS

Arts. 73—75 of the Code deal mainly with orthography of plant names.

The original spelling of a plant name is to be retained unless it was definitely wrong orthographically or can be shown to be a misprint when published. For example, the generic name *Clerodendrum* L. (Sp. Pl. 637. 1753 & in Gen. Pl. ed. 5. 285. 1754) was so spelled by Linnaeus and the spelling is not to be altered to *Clerodendron* as in most of the subsequent publications.

*Globba brachycarpa* Baker (in Hook. f. Fl. Brit. India 6 : 205. 1890) is a misprint (typographic error) for *Globba trachycarpa* Baker *Gluta benghas* L. (Mant. Pl. 293. 1771) is an orthographic error for *Gluta renghas* L. Linnaues used the vernacular name 'Renghas' as a specific epithet. Hence the binomial should be correctly cited as *Gluta renghas* L., as has been done by Engler (DC. Monogr. Phan. 4 : 225. 1883).

If two or more generic names are so similar leading to confusion they are to be treated as variants which are homonyms when they are based on differet types. For example, *Pleuripetalum* and *Pleuropetalum* are orthographic variants because only one can be used without serious risk of confusion. On the other hand, *Rubia* and *Rubus* are not likely to be confused and hence they are not treated as orthographic variants.

Similarly, for example, the epithets—*chinensis* and *sinensis*, *cey.*

*lonica* and *zeylanica*, *macrocarpon* and *macrocarpum* are treated as orthographic variants because they are likely to be confused when applied to related species within the same genus and to infraspecific epithets within the same species.

# 17

## HOW TO DESCRIBE A NEW GENUS/SPECIES

**New Genus :**

While describing a new genus, the name of a genus must be accompanied by a description or diagnosis of the genus followed by a description or reference to an earlier validly published description of the type species of that genus. Monotypic new genus based on a new species is validated by the provision of a combined generic and specific description or diagnosis. Botanists who are forming generic names should comply with the following suggestions:

I. (i) Generic name is a substantive in the singular number. It may be taken from any source whatever and even be composed arbitrarily.

(ii) The name of a genus should not coincide with a technical term used in morphology.

(iii) The name of a genus must be in one word. If it is formed by two words, that must be united by a hyphen.

(iv) Generic names should have a Latin termination.

(v) Names not readily adaptable to Latin language should be avoided.

(vi) Names which are very long or difficult to pronounce in Latin should be avoided.

(vii) Names should not be formed by combining words from different languages.

(viii) Adjectives used as nouns should be avoided.

(ix) A name similar to or derived from the epithet of one of the species of the taxon should not be used.

(x) It is recommended not to dedicate genera to person quite unconnected with botany or atleast with natural science.

(xi) It is to be given a feminine form to all personal generic names, whether they commemorate a man or a woman.

(xii) Names should not be formed by combining parts of two existing generic names because such names are likely to be confused with names of intergeneric hybrids.

(xiii) While naming after a person, the name should be formed in the following manner:

| *If the name of a person ends in :* | *To be added with it to form generic name :* |
|---|---|
| *Vowels* except *a* | *a*<br>*e.g. Santapaua* after Santapau |
| *a* | *ea*<br>*e.g. Collaea* after Colla |
| *ea* | *Correa* after Correa<br>(no letter is added) |
| *Consonants* except *er* | *ia*<br>*e.g. Adansonia* after Adanson |
| *er* | *a*<br>*e.g. Kernera* after Kerner |

| | |
|---|---|
| *us* (in latinized names) | The termination *us* is dropped before adding the suffix. *e.g. Dillenia* after Dillenius |

II. The type of a genus is a species, and hence the type-species of the new genus should be clearly indicated. If not already described, the type-species should be described (*vide* New Species).

III. All procedures have to be followed for valid publication (*vide* New Species).

**New Species :**

Once the taxonomic group of an undescribed taxon is fixed as a new species, the author has to follow certain taxonomic and nomenclatural procedures:

I. While choosing/forming a name (specific epithet) for the new species the following points should be observed :

(i) Specific epithets may be taken from any source whatever, and may even be composed arbitrarily.

(ii) The authors should avoid adoption of a name or epithet which has been previously used for a different species in the same genus.

(iii) Specific epithet must be a single word; if it consists of more than one, it must be united or hyphened. As far as possible it is better to avoid epithets formed of two or more hyphened words.

(iv) Specific epithet should not repeat the generic name.

(v) The epithets which are very much alike when used in the same genus should be avoided.

(vi) Very long epithets may not be chosen as it will be difficult to pronounce in Latin.

(vii) Specific epithets should not be formed by combining words from different languages.

(viii) Specific epithets which express a character common to all or nearly all the species of a genus may be avoided.

(ix) To avoid the names of little known or very restricted localities, unless species are quite local.

(x) The epithets must be with Latin terminations as far as possible.

(xi) An epithet named after a place is preferably an adjective and usually takes the termination—*ensis*,—(*a*) *nus*,—*ianus*, or—*icus*. e.g. *Jatropha tanjorensis* after Tanjore; *Andrographis nallamalayana* after Nallamalais; *Pogostemon travancoricus* after Travancore.

(xii) Names of men and women and also of countries and localities may be substantives in the genitive or adjectives.

(xiii) Specific epithet, when adjectival in form and not used as substantive, agrees grammatically with the generic name. *e.g. Brassica nigra*.

(xiv) While naming after a man, the name should be formed by adopting terminations in genitive cases as follows :

| *If the name of a person ends in :* | *To be added with it to form specific name :* |
|---|---|
| *Vowels* except a | *i* |
| | *e.g. Euphorbia santapaui* after Santapau |
| a | *e* |
| | *e.g. Euphorbia senguptae* after Sengupta |

| | |
|---|---|
| *Consonants* except *er* | *ii*<br>e.g. *Polygala jacobii* after Jacob |
| *er* | *i*<br>e.g. *Apama barberi* after Barber |

(xv) If the name of a person is already Latin or Greek, the appropriate Latin genitive should be used. *e.g. alexandri* from Alexander; *augusti* from Augustus; *linnaei* from Linnaeus.

(xvi) While naming after a person, the name can also be formed by adopting terminations in the adjectival form based on the gender of the generic names as follows :

| *If the name of a person ends in* : | *To be added with it to form specific name* : |
|---|---|
| *Vowels* and *er* | *an* (*a*); *an* (*us*) ; *an* (*um*).<br>e.g. *Crotalaria heyneana* after Heyne<br>*Lasianthus blumianus* after Blume<br>*Biophytum candolleanum* after de Candolle<br>*Verbena hasslerana* after Hassler |
| *Consonants* except *er* | *ian* (*a*); *ian* (*us*) ; *ian* (*um*). |

e.g. *Crotalaria grahamiana*
after Graham

*Calamus huegelianus*
after Huegel

*Ageratum houstonianum*
after Houston

II. Specific epithet should be followed by the author(s) name(s). Authors' names may be abbreviated, unless they are very short.

III. Rank of the taxon should be indicated following the author's name.

IV. A good, detailed description for the plant has to be prepared as it will be the original description. Description must be the recording of morphological characters of the group in an order. An excessively long description should not obscure essential points and at the same time short description will omit pertinent data. A description of a species should be fairly complete by stressing important selective characters. The style should be telegraphic, concise, consisting largely of nouns and adjectives in the nominative case. Strict grammatical sense should be maintained by using comma (,), semicolon (;), colon(:) and full stop (.). Description of the parts should follow a conventional sequence. It is usual to describe the habit and underground parts first. Then follows the description of leaves, inflorescence, flowers. The floral parts are described in an order—calyx, corolla,androecium and gynoecium. The fruits and seeds are described at the end. Measurement should be given in the metric system. The range of measurements of a part or average measurements are to be given.

V. Authors should give a full description or at least a diagnosis for the new species in Latin. The diagnosis should mention the points in which the new species differs from its allies. However, it is recommended to provide the diagnosis and description both in Latin for convenience.

VI. It is recommended that the description or diagnosis of parasitic plants should always be followed by indication of hosts.

VII. If possible authors should supply figures with details of structure as an aid to identification, the scale of the figures should be clearly indicated.

VIII. The description of a new species is followed by the indication of type with the particulars of its collections. The place where the type is permanently conserved also should be indicated. Indication of type should also be in Latin.

IX. The new species should be published in print in a well known widely circulated scientific journal/publication.

A new species described in Bull. Bot. Surv. India 6 :- 329 330. 1964 is reproduced here as a model :

# MODEL DESCRIPTION

## A NEW SPECIES OF *EUPHORBIA* FROM AGASTYAMALAI HILLS, MADRAS STATE

A. N. Henry

*Botanical Survey of India, Calcutta*

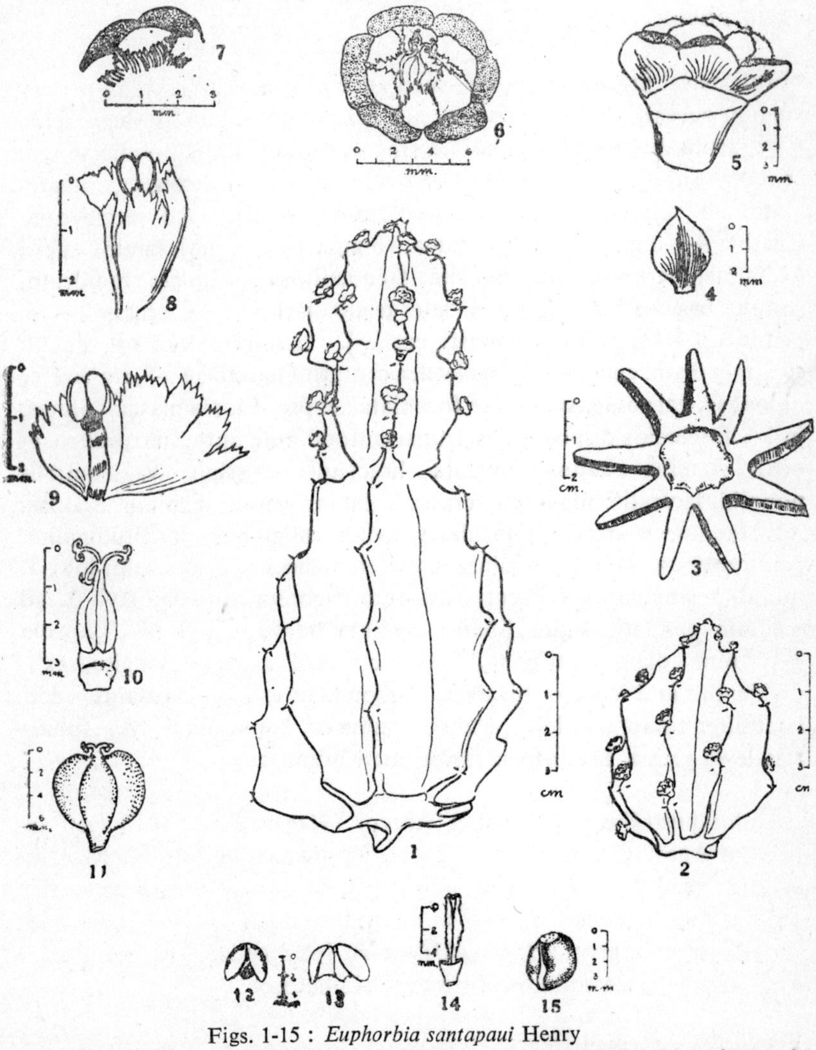

Figs. 1-15 : *Euphorbia santapaui* Henry

1 & 2. Portions of branches. 3. T. S. of branch. 4. Leaf. 5 & 6. Two views of cyathium. 7. Involucral lobe in between glands. 8. Male flower. 9. Bracteole spread out to expose stamen. 10. Female flower. 11. Capsule. 12 & 13. Two views of coccus, 14. Persistent axis of capsule. 15 Seed.

**Euphorbia santapaui** Henry, sp. nov.

Affinis *Euphorbiae antiquorum* Linn. a qua tamen differt ramis 5-9-alatis, cyathiis singulis (nec ternis), pedunculis abbreviatis, involucri lobis 5-8, staminibus penitus involutis bracteolis latissimis vaginantibus.

Frutices vel arbores carnosae, suco lacteo, 1.5-8 m altae; rami plures, sursum curvati, nodosi, 5-9-alati; alis gracilibus, lobulatis, bina spina brevi acuta divaricata stipulari ad singulos lobos. Folia 3-4mm longa, et lata, subsessilia, carnosa, late ovata vel suborbicularia, integra, glabra, ad apicem apiculata, ad basin obtusa, cito decidentia. Cyathia 4-9 mm diam., viridi-lutea, singula (nec terna), paulo supra axillam spinarum stipularium sita, frequentiora ad apices ramorum, suffulta bractea bina opposita late triangulari obtusa crassa rosea; pedunculi 2-4 mm longi, robusti, carnosi. Involucri lobi 5-8, singuli *ca* 2 mm diam. cucullati, irregulariter dentati. Glandulae 5-8, transverse oblongae, carnosae, spongiose foveolatae supra. Flores masculi plures, in 5-8 fasciculos dispositi (fasciculis tantum ante anthesim discernendos); stamina penitus involuta bracteolis latissimis vaginantibus nonnihil laceratis; filamenta crassa, glabra, supra medium nodosa; antherae luteae, erectae, biloculares, loculis subglobosis longitudinaliter dehiscentibus. Flores feminei breviter pedicellati; perianthii lobi 3, minuti, triangulares; subacuti; ovarium trigonum, roseum; styli 3, ad medium connati, supra liberi; stigmata biloba, papillosa. Capsula 7-10 mm diam., trigona, plano-convexa aspectu longitudinali, dividenda in 3 coccos bivalves, 1-seminales ex axi persistante; cocci lateraliter compressi, leves, glabri; semina *ca.* 3 mm diam. vix sphaerica levia glabra brunneola, raphe luteo-brunnea.

Holotypus (*Henry 17353A*) et isotypi (*Henry 17353 B-F*) lecti in collibus Agastyamalai dictis in Distr. Tirunelveli in ditione Madras ad altit. 1650 m die 27 augusti anni 1963; holotypus positus ad CAL; isotypi in herbario sectionis australis Bot. Surv. India ad Coimbatore. Paratypi (*Henry 16349 A-F*) lecti eodem in loco die 27 maii anni 1963 positi in herbario ad Coimbatore.

**Euphorbia santapaui** Henry, sp. nov.

Allied to *E. antiquorum* Linn., but differs by the branches 5-9-

winged, cyathia single (not in triads), in condensed peduncles, lobes of the involucre 5-8, and stamens quite enveloped in the very broad sheathing bracteoles.

Fleshy, glabrous shrubs or trees with milky juice, 1.5-8 m tall; branches numerous, curving upwards, jointed, 5-9-winged; wings thin, lobulate with one pair of short, sharp, divaricate stipular spines on each lobe. Leaves 3-4 mm long, as broad as long, subsessile, fleshy, broadly ovate or suborbicular, entire, glabrous, apiculate at apex, obtuse at base, soon deciduous. Cyathia 4-9 mm across, greenish yellow, single (not in triads), a little above the axils of stipular spines, more towards the apices of branches, each supported by two opposite, broadly triangular, obtuse, thick, pinkish bracts; peduncles 2-4 mm long, stout, fleshy. Lobes of the involucre 5-8, each *ca* 2 mm across, cucullate, irregularly toothed. Glands 5-8, transversely oblong, fleshy, spongily pitted above. Male flowers many, in 5-8 fascicles (only before anthesis the fascicles are easily recognisable); stamens quite enveloped in the very broad sheathing some what lacerate bracteoles; filaments thick, glabrous, jointed above the middle; anthers yellow, erect, 2-celled; the cells subglobose, opening longitudinally. Female flowers shortly pedicellate; perianth lobes 3, minute, triangular, subacute; ovary trigonous, pinkish; styles 3, connate to the middle, above that free; stigmas bilobed, papillose. Capsules 7-10 mm across, trigonous, plano-convex in the longitudinal aspect, separating into 3 bivalved, 1-seeded cocci, from a persistent axis; cocci laterally compressed, smooth, glabrous; seeds *ca* 3 mm across, almost spherical, smooth, glabrous, brownish, the raphe yellowish-brown. (Figs. 1-15).

The holotype of this species (*Henry 17353 A*) and isotypes (*Henry 17353 B-F*) were collected from Agastyamalai Hills in Tirunelveli district, Madras State at an altitude of about 1650 m on 27-8-1963; the holotype has been deposited in the Central National Herbarium, Botanical Survey of India, Howrah, and isotypes in the Southern Circle Herbarium, Botanical Survey of India, Coimbatore. The paratypes (*Henry 16349 A-F*), collected from the same locality on 27-5-1963 are also deposited in the Southern Circle Herbarium, Botanical Survey of India, Coimbatore, India.

SEE

The author wishes to dedicate this species to Rev. Fr. Dr. H. Santapau, S.J., F.N.I., Director, Botanical Survey of India, Calcutta, for his significant contributions to the taxonomy of Indian plants.

## ACKNOWLEDGEMENTS

Grateful thanks are due to the Director, Royal Botanic Gardens, Kew and Mr. A. R. Smith, also of Kew, for their valuable opinion on this species.

# NOMINA CONSERVANDA

The concept of *Nomina Conservanda* (abbreviated as *nom. cons.*) was introduced since the International Botanical Congress at Vienna, 1905. This system was introduced just to avoid disadvantageous changes in the names of families and genera by the strict application of the principle of priority. These names are conserved due to their very long usage in botanical publications and in the interest of nomenclatural stability.

Appendix II of the Code consists of a list of conserved families (Nomina Familiarum Conservanda). Many of the names included in the list are not in dispute and hence not in need of conservation, but their inclusion in the list is useful as a ready reference work giving correct names of families together with the type genus for each family.

Appendix III of the Code provides a list of conserved and rejected names of genera (Nomina Generica Conservanda et Rejicienda).

For example, the generic name *Desmodium* Desv. (J. Bot. Desv. ser. 2. 1 : 122. 1813) is conserved against the earlier names *Meibomia* Heist. ex Fabricius (Enum. 168. 1759) and *Pleurolobus* J. saint-Hilaire (Nouv. Bull. Soc. Philom. Paris 3 : 192. 1812), due to its very long usage in Botanical publications. The obsolete generic names (*Meibomia* and *Pleurolobus*) though earlier, are rejected.

The names in such lists are conserved against all earlier homonyms, and nomenclatural synonyms (different names based on the same type); and all names based on different types mentioned in the list.

Any proposal of names for conservation has to be submitted to the General Committee with a clear statement of the cases both for and against its conservation. This will be referred for the examination to the Committees for the various taxonomic groups. The proposals of conservation have to be approved by the General Committee of the concerned taxonomic group and then by the next International Botanical Congress by voting system.

When a conserved name competes with one or more other names based on different types and against which it is not conserved, the earliest of the competing name is adopted against the conserved one.

*e.g. Nasturtium* R.Br. (1812) was conserved only in the restricted sense for a monotypic genus based on *N. officinale* R. Br. If it is reunited with *Rorippa* Scop. (1760), it must bear the prior name *Rorippa*.

Van Steenis (Taxon 13: 155-156. 1964) suggested the following as guiding principles in the matter of conservation : (1). That the spirit of the Code is to promote stability of nomenclature, and that this supersedes the principle of priority, where possible, reasonable and desirable. (2). That conservation of generic names, is intended to stabilize nomenclature, irrespective of size or economic importance of taxa. (3). That consequentely those names must be conserved which are longest in general use, and that obsolete names, whatever the scientific merits of their authors, should not replace names which are in current usage.

**Nomina Specifica Conservanda**

Conservation of specific names is not legislated in the "Code". Strict application of the principle of priority has caused many changes in the names of species. Though there has been hot discussions for years since Cambridge Congress, 1930, no general agreement has been made for conservation of specific names. Fosberg (Taxon 13 : 177-178. 1964) strongly objected to *nomina specifica conservanda* on the plea that such a list would render the Code almost unusable. Lanjouw (Taxon 12 : 305-309. 1963) suggested as the solution of the problem that only important plants should be considered. Here it is very difficult to decide what are important plants. De Wit (Taxon 13: 86. 1964) states that many plants which were important economically have now lost their economic importance wholly or partly. De Wit (*loc. cit.* 88) states "I wish to stress the fact that the problem of stabilising specific names of economic (important) species, has been brought up again and again. The Congresses have repeatedly rejected the proposals, not by a whim or at the spur of the moment but after repeated discussions, and close scrutiny. It was certainly not a light-hearted refusal. A special committee could not reach a solution. Why insist on bringing up a problem that has been rejected so consistently and so frequently". The reason for the recurrence of the problem seems to be that some users of plant names, *e.g.* foresters, agricultural botanists, horticulturists, etc., have again and again manifested that they are tired of the constant changing of names.

# 19

## SOME IMPORTANT DATES IN BOTANICAL NOMENCLATURE

1.5.1753 Beginning for valid publication of names for plants of Spermatophyta and Pteridophyta (Linnaeus, *Species Plantarum* ed. 1).

1.5.1753 Beginning for valid publication of names for plants of Sphagnaceae and Hepaticae (Linnaeus, *Species Plantarum* ed. 1).

1.5.1753 Beginning for valid publication of names for plants of Lichens (Linnaeus, *Species Plantarum* ed. 1).

1.5.1753 Beginning for valid publication of names for plants of Algae (Linnaeus, *Species Plantarum* ed. 1).

Exceptions : Nostocaceae, Homocysteae, 1.1.1892.

Nostocaceae, Heterocysteae, 1.1.1886.

Desmidiaceae, 1.1.1848.

Oedogoniaceae, 1.1.1900.

1.5.1753 Beginning for valid publication of names for plants of Myxomycetes (Linnaeus, *Species Plantarum* ed 1).

1.1.1801 Beginning for valid publication of names for plants of Musci except Sphagnaceae (Hedwig, *Species Muscorum*).

31.12.1801 Beginning for valid publication of names for plants of Fungi : Uredinales, Ustilaginales and Gasteromycetes (Persoon, *Synopsis Methodica Fungorum*).

31.12.1820 Beginning for valid publication of names for all fossil plants (Sternberg, *Flora der Vorwelt, Versuch* 1 : 1-24. t. 1-13).

1.1.1821 Beginning for valid publication of names for plants of FungiCrteri (Fries, *Systema Mycologicum* Vol. 1).

1.1.1908 The name of a genus, species or an infraspecific taxon published before this date is valid even if it is accompanied only by an illustration with analysis showing essential characters.

1.1.1935 A name of a new taxon of plants (except algae and fossils) published on or after this date must be accompanied by a Latin description or diagnosis or by a reference to a previously and effectively published Latin description or diagnosis of the taxon, for a valid publication.

1.1.1953 Publication of a new name by indelible autograph before this date, constitutes effective publication.

1.1.1953 Publication of a new name in tradesmen's catalogues ot non-scientific newspapers on or after this date does not constitute effective publication.

1.1.1953 The distribution of printed matter accompanying exsiccata before this date constitutes effective publication.

1.1.1953 A new combination, published on or after this date, for a previously and validly published name is not valid unless its basionym is clearly indicated with its full citation.

1.1.1953 A proposed new name (*nomen novum*) published on or after this date, for a previously and validly published name is not valid unless its replaced synonym is clearly indicated with full citation of its original publication.

1.1.1953 A new name or combination published on or after this date without clear indication of the rank of the taxon concerned is not validly published.

1.1.1958 A name of a new taxon of recent algae of specific or lower rank published on or after this date must be accompanied by an illustration or figure showing distinctive morphological features.

1.1.1958 Publication of the name of a new taxon of the rank of family or below on or after this date is valid only when the nomenclatural type is indicated.

1.1.1958 A name of a new taxon of recent algae published on or after this date must be accompanied by a Latin description or diagnosis or by a reference to a previously and effectively published Latin description or diagnosis of the taxon.

1.1.1973 A name published on or after this date for which the various conditions for valid publication are not simultaneously fulfilled is not validly published unless a full or direct reference is given to the places where these requirements were previously fulfilled.

# 20

## INTERNATIONAL ASSOCIATION FOR PLANT TAXONOMY (IAPT)

IAPT is an international body which maintains the International Bureau for Plant taxonomy and Nomenclature, with its headquarters in Utrecht, Netherlands. "Taxon" is the journal of the above association. IAPT edits and publishes the successive editions of the 'International Code of Botanical Nomenclature' in conformity to the decisions adopted by the various International Botanical Congresses held at an interval of about five years. Besides the Code, its series entitled '*Regnum Vegetabile*' includes several valuable publications in Plant Taxonomy like the Index Herbariorum, International Directory of Specialists in Plant Taxonomy and Taxonomic Literature.

Any individual or Institution can become a member of IAPT. Members are entitled to vote at the Nomenclature section meetings at International Botanical Congresses.

Permanent Nomenclature Committees are established under the auspices of the IAPT; members of these Committees are elected in the International Botanical Congresses :

1. General Committee—Composed of the Secretaries of other committees, the Rapporteur-general, the President and Secretary of IAPT, and at-least five members to be appointed by the Nomenclature Section.

2. Committee for Spermatophyta.

3. Committee for Pteridophyta.

4. Committee for Bryophyta.

5. Committee for Fungi and Lichens.

6. Committee for Algae.

7. Committee for Hybrids.

8. Committee for Fossil plants.

9. Editorial Committee—Responsible for the preparation and publication of the "Code" in conformity to the decisions arrived at the International Botanical Congress.

The Bureau of Nomenclature of the International Botanical congress consists of :

1. The President of Nomenclature Section.

2. The Recorder.

3. The Rapporteur-general.

4. Vice-Rapporteur.

The Rapporteur-general is in charge with the presentation of nomenclature proposals to the International Botanical Congress.

The voting on nomenclature proposals is of two kinds :

1. **Preliminary guiding mail vote :**

The members of the IAPT, authors of proposals and the members of Nomenclature Committees are eligible.

2. **Final and binding vote at the Nomenclature Section of the International Congress :**

All officially enrolled members of the section and official delegates or vice-delegates of the institutes appearing on a list drawn up by the Bureau of Nomenclature of the International Botanical Congress are qualified for final votings at the sessions.

21

# TEACHING OF BOTANICAL NOMENCLATURE THROUGH PRACTICAL APPLICATION

Systematic Botany or Systematics embraces the whole field of systematic work and is broadly divisible into two parts : *Taxonomy* deals with the placement of an individual plant into a taxonomic group or taxon, and the assignment of the taxon into the general system of classification which is, of course, phylogenetic in nature; and *Nomenclature* deals with the determination or selection of the correct name to be applied to a known taxon in conformity with the International Code of Botanical Nomenclature. Systematic work is correctly and fully done only when both these steps or stages (*viz.* taxonomic and nomenclatural) are properly carried out. Nomenclature thus forms an inseparable and important part of Systematic Botany. Undoubtedly, nomenclature serves taxonomy.

In various floras of India published up to the early part of 20th century, much attention was not paid to the selection of the correct names of plants. Relevant synonymy was also invariably omitted. These have caused much confusion in the identity and nomeclature of several common Indian plants. A break-through in floristic research in India was noticed in 1953 when Santapau published his "Flora of Khandala on the Western Ghats of India" adopting the correct identity and nomenclature of the plants treated therein and also by citing relevant synonymy. Several regional/district floras of India published since then, have followed suit. The plant names given in these floras no doubt vary considerably from the old floras. The majority of the recent name changes of Indian plants are due to strict application of the International Code of Botanical Nomenclature, while others are either due to the better understanding of the identity of the plant or even to the proper judgement of the taxonomic status of the taxa. Hence identity and nomenclature are equally important and they should go side by side in fixing the correct names of plants.

Santapau (*Sci. & Cult.* 31 : 467. 1965) stressed that the "Code" should be included in the curriculum of atleast such post-graduate students who take up any of the branches of plant systematics for their special study. It is gratifying to note that in recent times it has gained increased recognition and this is reflected by a large number of colleges and universities in India that include it in their syllabi. Normally taxonomic part is taught at length by lectures, laboratory work and on field excursions; but nomenclature is usually covered briefly in a few lectures that are mostly historical in view point, as opposed to practical. These lectures, no doubt, are of value, but the student does not gain a detailed knowledge of the laws of the Code, and this creates difficulty for him to follow the nomenclatural/taxonomic synonymy given in recent floras/monographs and to arrive at the correct names of plants. While naming the plants, he still uses the incorrect names given in the old out-dated floras. Hence a very effective way of training botanical students in nomenclature is by the *case method* of the International Code of Botanical Nomenclature. Harold St. John, as early as in 1958, stressed this aspect in his "Nomenclature of Plants". Each student should be able to investigate and evaluate the validity of the publication cited, search for synoymy in classical books/literature and for pertinent facts such as basionyms, homonyms, tautonyms and to understand the terms often used such as *comb. nov.* and *nom. nov.*

Solutions including process and investigation of three sample nomenclature cases are given below :

(i) *Ophioxylon serpentinum* L. Sp. Pl. 1043. 1753.

*Ophioxylon trifoliatum* Gaertn. Sem. Pl. 2: 123. 1791.

*Rauvolfia serpentina* (L.) Benth. ex Kurz, For. Fl. Burma 2: 171. 1877.

The earliest of these names is *Ophioxylon serpentinum* which is found in Linnaeus' *Species Plantarum*—1753. There on page 1043 Linnaeus validly published this name (according to Arts. 32-45 of the International Code of Botanical Nomenclature). Further, according to Art. 13, valid publication of names for Spermatophyta and Pteridophyta begins from 1st May, 1753 (Linnaeus, *Species Plantaram* ed. (1). In Linnaeus' Species Plantarum the placing of the epithet in the margin opposite the name of the genus clearly indicates the combination intended (Art. 33). Bentham (in Genera Plantarum 2: 697. 1876) appears to have been the first in uniting *Ophioxylon* L. (Sp. Pl. 1043. 1753; Gen. Pl. ed. 5. 467. 1754) and *Rauvolfia* L. (Sp. Pl 208. 1753; Gen. Pl. ed. 5. 98. 1754), after adequate comprehension of the generic characteristics of bot h the genera. The issue is of course a taxonomic one. He adopted the name *Rauvolfi*a for the combined genus and this name is accordingly to be retained (Art. 57.2). Bentham did not really effect the transfer of the species *Ophioxylon serpentinum* L. to *Rauvolfia*.

*Rauvolfia serpentina* (L.) Benth. ex Kurz was a combination based on the oldest epithet-bringing synonym (basionym)—*Ophioxylon serpentinum* L. (Art. 33.2). When a speces is transferred to another genus but retains its epithet the author of the basionym (who published this as a legitimate name) must be cited in parantheses, followed by the author who effected the combination (Art. 49). Kurz in his *Forest Flora Burma* 2: 171. 1877 first validly published the combination by directly giving reference to the basionym, but ascribed it to Bentham. According to Recommendation 46C. I, the correct author citation is the name of the publishing author (Kurz), but the name of the person followed by the connecting word *ex* may be inserted before the name of the lublishing author, if desired (*i.e.* Benth. ex Kurz).

Another question of some concern is the orthography of the generic name. Plumier followed by Linnaeus consistently used the Latin version of Rauwolf's name and named the genus as *Rauvolfia*. But Willdenow in his Species Plantarum and following his several others including authors of Indian floras spelt the generic name as *Rauwolfia*. However, according to Art. 73, the original spelling of Linnaeus (intentional latinisation of Rauwolf's name) *viz*. *Rauvolfia* is to be preserved.

In 1791, Gaertner (Fruct. Sem. Pl. 2: 123) validly published the name—*Ophioxylon trifoliatum*. However, this name became superfluous (Art. 63), as Gaertner's plant already had an earlier, validly published name—*O. serpentinum* L. (1753).

Hence, the correct name of 'Sarpagandha' is *Rauvolfia serpentina* (L.) Benth. ex Kurz.

(ii) Another case involving both identity and nomenclature is discussed below :

**Entada pursaetha** DC. Prodr. 2 : 425. 1825.

*Mimosa entada* L. Sp. Pl. 518. 1753.

*Entada rheedii* Spreng. Syst. 2: 325. 1825.

*Entada monostachya* DC. Prodr. 2: 425. 1825.

*Entada scandens* auct. non Benth. 1841; Gamble, Fl. Pres. Madras 417. 1919.

Most of the earlier Indian floras report the occurrence of *Entada scandens* (L.) Benth. in India. However, recent critical studies undertaken by way of 'type method' have revealed that the true *Entada scandens* (L.) Benth. which is synonymous to *Entada phaseoloides* (L.) Merrill does not occur in India, but found only in the Amboina in the Moluccas; and the correct identity of the common *Entada* occurring in India should be *Entada pursaetha* DC.

Now investigation of the nomenclature case reveals :

The earliest of these names is *Mimosa entada* which was validly published in Linnaeus' Species Plantarum p. 518. 1753. Augustin de Candolle (1825) while transferring this species to the genus *Entada*, could not retain the specific epithet "*entada*" as the resulting binary name "*Entada entada*" is a tautonym which is inadmissible according to Art. 23. Hence he proposed a new name *Entada monostachya* DC. (in his Prodr. 2: 425. 1825). The three competing names for this species in the genus *Entada* viz. *E. pursaetha* DC., *E. rheedii* Spreng.

and *E. monostachya* DC. all date from 1825. Brenan (Kew Bull. 1955: 164. 1955) appears to have been the first to unite all the above three species; he adopted the name *Entada pursaetha* DC. for the combined species, and this name is accordingly to be retained (Art. 57. 2).

Now regarding the citation of the misapplied name: According to Recommendation 50D. 1, the name *E. scandens* as a misidentification should not be included in the synonymy of *E. pursaetha* but added after it. Further, the misapplied name (*i.e. Entada scandens*) should be indicated by the words *auct. non* followed by the name of the original author (Benth.) and the bibliographical reference of the misidentification, *i.e.* reference to Gamble, Fl. Pres. Madras or any other floras as the case may be which have misidentified the plant.

The correct identity and nomenclature of the common Indian species of *Entada* is, therefore, determined as *Entada pursaetha* DC.

(iii) Another nomenclature case involving the proper judgement of the taxonomic status of two genera is given below :

In most of the older floras, the genera *Abelmoschus* Medicus, Malv. 46. 1787 and *Hibiscus* L. (Sp. Pl. 693. 1753; Gen. Pl. ed 5. 310. 1754) are treated as congeneric (*i.e.* synonymous). However, K. Schumann (in Engler & Prantl, Nat. Pflanzenfam. 3(6): 47. 1895) and following him several monographers treated them as distinct based mostly on the nature of the calyx: spathaceous, irregularly 2 to 3-lobed and caducous in *Abelmoschus*; and campanulate, cupular, regularly 5-lobed or truncate with minute teeth, and persistent in *Hibiscus*.

Consequently several species of *Hibiscus* including *H. esculentus* L. were transferred to genus *Abelmoschus* :

**Abelmoschus esculentus** (L.) Moench, Meth. Pl. 617. 1794.

*Hibiscus esculentus* L. Sp. Pl. 696. 1753.

*Hibiscus longifolius* Willd. Sp. Pl. 3 : 827. 1800.

*Abelmoschus esculentus* (L.) Moench was a combination based on the oldest epithet-bringing synonym (basionym)—*Hibiscus esculentus* L. (Art. 33.2). Moench in his *Methodus Plantas* (1794) first validly published the combination by directly giving reference to the basionym. The author of the basionym is cited in parantheses, followed by the author who effected the combination (Art. 49).

In 1800, Willdenow (Sp. Pl. 3 : 827) validly published the name *Hibiscus longifolius*. However, this name became superfluous (Art. 63) as Willdenow's plant already had a prior validly published name-*H. esculentus* L. (1753).

Hence the correct name of 'bhindi' is *Abelmoschus esculentus* (L.) Moench.

The solutions of even these simple nomenclature cases bring the student in contact with several of the fundamental botanical publications. The correct interpretation depends on an understanding of the principles of priority, synonymy, regulation governing the binomial system, and other concerned Articles and Recommendations of the International Code of Botanical Nomenclature. Several other cases can be digested and solved in a similar way and certainly the study will aid in giving the student a sounder training in Botany. Only after investigation and evaluation of a few cases, he evinces interest in comparing the old and recent floras for name changes and in course of time will be fascinated to fix for himself the correct identity and nomenclature of the common local plants in conformity with the rules of the International Code of Botanical Nomenclature.

A few more cases which will be useful to illustrate the application of the laws of nomenclature are given below ; the solutions of them can be worked out by the students.

(i) **Erthroxylum moonii** Hochr. in Bull. L'inst. Bot. Buitenzorg 22 : 54. 1905.

*E. acuminatum* (Arn.) Walp. Rep. 1 : 407. 1842, non Ruiz & Pav. 1802.

*Sethia acuminata* Arn. in Nov. Act. Acad. Nat. Cur. 18 (1) : 324. 1836.

*Erythroxylum lucidum* Moon [Cat. Pl. Ceylon 36. 1824 *nom. nud.*] ex Hook. f. Fl. Brit. India 1 : 415. 1874, non HBK. 1821.

(ii) **Saprosma foetens** (Wight) K. Schum. in Engl. & Prantl, Nat. Pflanzenfam. IV (4) : 122. 1891.

*Lasianthus foetens* Wight in Calc. J. Nat. Hist. 6: 517. 1846.

*Saprosma ceylanica* (Gardn.) Bedd. For. Man. 134/12. 1872.

*Dysodidendron ceylanicum* Gardn. in Calc. J. Nat. Hist. 7 : 2. 1847.

(iii) **Sesbania** Scop. Intr. 308. 1777, *nom. cons.*

*Sesban* Adans. Fam. 2 : 327. & 604. 1763.

*Agati* Adans. Fam. 2 : 326 & 513. 1763.

(iv) **Baeolepis** Decne. ex Moq. in DC. Prodr. 13 (2) : 216. 1849. *Brachylepis* Wight & Arn. in Wight, Contrib. Bot. Ind. 63. 1834, non C.A. Mey. 1829, nec Hook. & Arn. 1834.

(v) **Jatropha heynei** Balak. in Bull. Bot. Surv. India 3: 40. 1961.

*J. heterophylla* Heyne ex Hook. f. Fl. Brit. India 5 : 382. 1887 (non Steudel 1840).

(vi) **Boerhavia chinensis** (Burm. f.) Druce in Rep. Bot. Exch. Cl. Brit. Isles 1913 (3) : 415. 1914.

*Valeriana chinensis* Burm. f. Fl. Ind. 15. t. 6. f. 3. 1768.

*Boerhavia repanda* Willd. Sp. Pl. 1 : 22. 1797.

(vii) **Cadaba fruticosa** (L.) Druce in Rep. Bot. Exch. Cl. Brit. Isles 1913 (3) : 415. 1914.

*Cleome fruticosa* L. Sp. Pl. 671. 1753.

*Cadaba farinosa* Forsk. Fl. Aegypt.—Arab. 68. 1775.

*Cadaba indica* Lamk. Encycl. 1 : 544. 1785.

(viii) **Cleome gynandra** L. Sp. Pl. 671. 1753.

*C. pentaphylla* L. Sp. Pl. 983. 1762.

*Gynandropsis pentaphylla* (L.) DC. Prodr. 1: 238. 1824.

*G. gynandra* (L.) Briq. in Ann. Cons. Jard. Bot. Geneve 17 : 382. 1914.

(ix) **Pterolobium hexapetalum** (Roth) Sant. & Wagh in Bull. Bot. Surv. India 5 : 108. 1963.

*Reichardia hexapetala* Roth, Nov. Pl. Sp. 210. 1821.

*Pterolobium indicum* A. Rich. Tent. Fl. Abyss. 1: 247. 1842.

(x) **Trapa natans** L. var. **bispinosa** (Roxb.) Makino in Iinuma, Somoku Dzusetzii (ed. 3) 1 : 137. 1907.

*T. bispinosa* Roxb. Pl. Cor. 3: t. 234. 1815.

## EXERCISE

Q. *Theka* Adans. Fam. 2 : 445. 1763.

*Tectona* L. f. Suppl. 151. 1781. *nom. cons,*

Both these genera are synonymous. Select the correct name giving reasons.

A. *Tectona* L. f. (1781) is the correct name as it is conserved against *Theka* Adans., though the latter name was published earlier (1763). According to Article 11 of the Code, for any taxon from family to genus inclusive, the correct name is the earliest legitimate one with the same rank except in cases of limitation of priority by conservation. Article 14 of the Code gives detailed prescriptions for conservation of names; also the Code provides in Appendix III, a list of names of conserved and rejected genera. Conservation of generic names is primarily intended to stabilize nomenclature, and to avoid disadvantageous changes in the nomenclature of genera by the strict application of the principle of priority. The generic name *Tectona* L. f. (1781) is conserved against the earlier name *Theka* Adans. (1763) due to its very long usage in botanical publications ; the obsolete generic name *Theka*, though earlier, is rejected.

Q. **Pilea melastomoides** (Poir.) Wedd. in Ann. Sci. Nat. (Ser. 4) 1: 186. 1854. *Urtica melastomoides* Poir. in Lamk. Encycl. Suppl. 4 : 223. 1816. *Pilea trinervia* Wight, Icon. t. 1973. 1853.

Select the basionym, and state why *Pilea trinervia* is relegated to synonymy.

**A.** In nomenclature citations *Urtica melastomoides* Poir. (1816) is the basionym or the epithet-bringing synonym of *Pilea melastomoides* (Poir.) Wedd. *Urtica melastomoides* was first published legitimately by Poirret, and when transferred to the genus *Pilea*, the specific epithet '*melastomoides*' must be retained, and the author of the basionym (Poirret) must be cited in parantheses, followed by the name of the author who effected the alteration (the author of the new name *i.e. Weddell*).

*Pilea trinervia* Wight is a taxonomic synonym of *Pilea melastomoides* (Poir.) Wedd. as it is based on a different type. The binomial, *Pilea trinervia* Wight (1853) became superfluous, according to Art. 63 of the Code as this plant already had an earlier published name—*Urtica melastomoides* Wedd. (1816).

**Q.** Explain *Elatostema lineolatum* Wight var. *lineolatum.* Note the final (varietal) epithet repeats unchanged of the specific epithet.

**A.** *Elatostema lineolatum* Wight var. *lineolatum* is an autonym (*i.e.*) a name automatically established for a taxon, merely as a consequence of the proposal of another taxon. The recognition of a variety of *Elatostema lineolatum* that does not contain its type, automatically establishes a variety that does contain the type. Hence while indicating the typical variety that contains the nomenclatural type of the species, the specific epithet—'*lineolatum*' has been repeated unaltered but without citation of author's name (Wight) : *Elatostema lineolatum* Wight var. *lineolatum*.

**Q.** *Dregea angustifolia* (Hook. f.) Santapau & Irani in Bot. Mem. Univ. Bombay 4 : 41. 1960.

*D. volubilis* var. *angustifolia* Hook. f. Fl. Brit. India 4 : 47. 1883.

*Marsdenia lanceolata* Cooke, Fl. Pres. Bombay 2 : 167. 1904.

*Dregea lanceolata* (Cooke) Santapau & Wagh in Bull. bot. Surv. India 5 : 106. 1964.

What is the correct name of the above taxon when treated as a species of *Dregea* ?

A. *Dregea lanceolata* (Cooke) Santapau & Wagh is the correct name of the taxon when it is treated as a species of *Dregea*. According to Article 60, when the rank of a genus or infrageneric taxon is changed, the correct name or epithet is the earliest legitimate one available in the new rank. In no case does a name or an epithet have priority outside its own rank. In the rank of a species, *Marsdenia lanceolata* Cooke (1904) is the earliest legitimate name, though this taxon was originally treated as a variety of *Dregea volubilis* viz. *D. volubilis* var. *angustifolia* Hook. f. (1883).

*Dregea lanceolata* (Cooke) Santapau & Wagh (1964) is the correct name, since Santapau & Wagh have retained the specific epithet '*lanceolata*' in *Dregea*. The varietal epithet '*angustifolia*' has no priority outside its own rank and hence the combination *Dregea angustifolia* (Hook. f.) Santapau & Irani (1960) is illegitimate and must be rejected.

Q. Are the following family names correct ? Explain. Aroideae, Annonaceae, Gramineae.

A. The name of a family is formed by adding the termination—*aceae* to the stem of a legitimate genus (type genus). The family name Aroideae is incorrect. *Arum* is the type genus of the family and hence Araceae is the correct name of the family. The family name Annonaceae is correct as it is based on the type genus *Annona*. The usage of the family name Gramineae, though not derived from the stem of a generic name, is treated as validly published because of its long usage. The Code has, however, authorised the usage of the alternative name Poaceae based on the type genus *Poa*.

Q. *Strychnos nux-vomica* L. Is it a binary combination ?

A. According to Article 33, the name of a species is a binary combination consisting of the name of the genus followed by a single specific epithet. If an epithet consists of two or more words, these must be united or hyphened. An epithet not so joined when originally published is not to be rejected but, when used must be united or hyphened. Though in this case, Linnaeus originally published the specific epithet as 'Nux Vomica', when used it must be united as *nux-vomica.* Hence *Strychnos nux-vomica* L. is a binary combination.

Q. *Cajanus cajan* (L.) Millsp. Is the binomial a tautonym ?

A. *Cajanus cajan* is not a tautonym as the specific epithet does not exactly repeat the generic name; hence the binomial is legitimate.

Q. When the three genera—*Sloanea* L. (1753), *Echinocarpus* Blume (1825) and *Phoenicosperma* Miq. (1865) are united, which name should be adopted for the resulting genus ? Explain.

A. *Sloanea* L. (1753) is the oldest of these three generic names and hence should be adopted for the resulting genus.

Q. Bentham (1876) first united the genera *Ophioxylon* L. (1753) and *Rauvolfia* L. (1753) adopting the name *Rauvolfia* for the combined genus. Was his choice correct ?

A. Yes. According to Art. 57.2, the author who first unites taxa bearing names or epithets of the same date must choose one of them, and his choice must be followed. Bentham's choice of the name *Rauvolfia* for the combined genus is correct and must be retained.

Q. *Cyperus iria* var. *rectangularis* Kukenth. in Engler, Pfreich. 101 : 152. 1935.

*Cyperus rectangularis* (Kukenth.) Bennet in Indian For. 95 : 692. 1969.

*Cyperus alulatus* Kern in Reinwardtia 1 : 463. 1952.
What is the correct name of the taxon when treated as a species ?

A. *Cyperus alulatus* Kern is the correct name of the taxon when it is treated as a species. In no case does a name or an epithet have priority outside its own rank (Art. 60). The varietal epithet—*rectangularis* though published earlier, has no priority outside its own rank and hence the combination *Cyperus rectangularis* (Kukenth.) Bennet (1969) is illegitimate and must be rejected.

Q. *Mahonia* Nutt. 1818, *nom. cons.*

*Berberis* L. 1753.

If these two genera are united what will be the correct name for the combined genus ? Explain.

A. When *Mahonia* Nutt. (1818) *nom. cons.* is united with *Berberis* L. (1753), the combined genus will bear the prior name *Berberis* L. although *Mahonia* is conserved and *Berberis* is not; because according to Art. 14.3, when a conserved name competes with one or more other names based on different types and against which it is not explicitly conserved, the earliest of the competing names is adopted.

Q. *Pyrus malus* L. Sp. Pl. 479. 1753.

*Malus pumila* Mill. Gard. Dict. ed. VIII. n. 3. 1768.

*Malus malus* (L.) Britton in Britton & Brown, Ill. Fl. Ed. 2, 2 : 290. 1913.

All the three names are nomenclatural synonyms (*i.e.*) based on the same type. What is the correct name if the species is treated under the genus *Malus* ? Explain.

A. The earliest of these names is *Pyrus malus* L. (1753). Miller (1768) while transferring this species to the genus *Malus* could not retain the specific epithet '*malus*' as the resulting binary name '*Malus malus*' is a tautonym which is inadmissible according to Art. 23.4; hence he correctly proposed a new name *Malus pumila* Mill. (1768). The subsequent publication of the tautonym *Malus malus* (L.) Britton (1913) is illegitimate. Hence the correct name of the species when treated under the genus *Malus* is *M. pumila* Mill.

Q. The family name Ebenaceae Giirke, *nom. cons.* is based on the stem of the illegitimate generic name *Ebenus* O. Ktze., non L.; state whether the name Ebenaceae is legitimate or not ?

A. The family name Ebenaceae is legitimate because it is conserved. According to Art. 18.3, a name of a family based on the stem of an illegitimate generic name is illegitimate unless conserved.

Q. "**Indobanalia** Henry & Roy, *nom. nov.* *Banalia* Moq. in DC. Prodr. 13(2): 278. 1849, *non.* Rafin. Autikon Botanikon 50. 1840." The new name (*nom. nov.*) *Indobanalia* was published by & Roy in Bull. Bot. Surv. India 10 : 274. 1968 (1969). Whether it was validly published or not ? Why the new name was published ?

A. For valid publication, a name of taxon should comply with provisions of Arts. 32 - 45 :

The new name *Indobanalia* Henry & Roy was effectively published in Bull. bot. Surv. India. Also the replaced synonym *Banalia* Moq. was clearly indicated and a full and direct reference given to its author and original publication. Hence *Indobanalia* Henry & Roy was validly published.

*Banalia* Moq. (1849) is a later homonym of *Banalia* Rafin. (1840). *Banalia* Moq. is therefore rejected (Art. 64.1), as was done by Henry & Roy in (Bull. Bot. Surv. India 10 : 274. 1968 (1969), who renamed it *Indobanalia*.

Q. Imagine that the genus *Euphorbia* L. *sensu lato* (in the wider sense) is divided into genera like *Chamaesyce* Gray, *Poinsettia* Grah. and *Tithymalus* Gaertn.; state whether the above procedure is correct or not ?

A. This procedure is contrary to the rules of Nomenclature.

According to Art. 52.1, when a genus is divided into two or more genera, the generic name must be retained for one of them. Hence, *Euphorbia* L. *sensu stricto* (in the restricted sense) must be kept for one of the genera which includes the type of *Euphorbia* L.

**Q.** *Tabernaemontana gamblei* Subr. & Henry in Bull. Bot. Surv. India 12 : 5. 1970 (1972).

*Ervatamia caudata* Gamble in Kew Bull. 1921 : 310. 1921, non *Tabernaemontana caudata* Merr. 1909.

After studying the synonymy state why a new name *Tabernaemontana gamblei* Subr. & Henry was published when *Ervatamia caudata* Gamble was transferred to the genus *Tabernaemontana.*

**A.** *Ervatamia caudata* Gamble (1921) when transferred to the genus *Tabernaemontana* must be called *Tabernaemontana gamblei* Subr. & Henry (1972), because of the existence of the name *Tabernaemontana caudata* Merr. (1909) referring to a different species. In other words, since the epithet '*caudata*' was preoccupied in the genus *Tabernaemontana*, a new name *T. gamblei* Subr. & Henry (1972) was correctly published.

**Q.** *Rauvolfia serpentina* (L.) Benth. ex Kurz, For. Fl. Burma 2 : 171. 1877.

*Basionym* : *Ophioxylon serpentinum* L. Sp. P1. 1043. 1753.

Explain the usage of L. in parantheses and *ex* in the author citation *i.e.* (L.) Benth. ex Kurz.

**A.** *Rauvolfia serpentina* (L.) Benth. ex Kurz was a combination based on the basionym (epithet-bringing synonym) *Ophioxylon serpentinum* L. (Art. 33.2). When a species is transferred to another genus but retains its epithet, the author of the basionym must be cited in parantheses, followed by the author who effected the combination (Art. 49). Kurz, For. Fl. Burma 2 : 171. 1877 validly published the combination by directly giving reference to the basionym, but ascribed it to Bentham. According to Recommendation 46C I, the correct author citation is the name of the publishing author (*i. e.* Kurz), but the name of the other person followed by the connecting word *ex* may be inserted before the name of the publishing author, if desired (*i. e.*) *Rauvolfia serpentina* (L.) Benth. ex Kurz.

**Q.** *Croton roxburghii* Balak. in Bull. bot. Surv. India 3 : 39. 1961. *C. oblongifolius* Roxb. Fl. Ind. 3 : 685. 1832 non Delile 1814, nec Siebr. ex Spreng. 1826.

Is the new name *Croton roxburghii* Balak. Correct ? Also explain the usage of 'non' and 'nec' in the citation.

**A.** *Croton oblongifolius* Roxb. (1832) is a later homonym of *Croton oblongifolius* Delile (1814). Another later homonym is *Croton oblongifolius* Siebr. ex Spreng. (1826). When a later homonym (which is illegitimate) is cited in synonymy, the citation should be followed by the name of the author of the earlier homonym preceded by the word *non*, with the date of publication added (*i.e.* non Delile 1814). Any other later homonym is also cited, preceded by the word *nec* (*i.e.* nec Siebr. ex Spreng. 1826).

The later homonym, *Croton oblongifolius* Roxb. (1832) being illegitimate is rejected (Art. 64), and is replaced correctly by the new name *Croton roxburghii* Balak. (1961).

**Q.** *Sarcandra grandifolia* (Miq.) Subr. & Henry in Bull. bot. Surv. India 12 : 5. 1970 (1972).

*Chloranthus grandifolius* Miq. Fl. Ind. Bat. 1 (1) : 802. 1856.

*Sarcandra irvingbaileyi* Swamy in Proc. nat. Inst. Sci. India B 19 : 303. 1953.

*Sarcandra chloranthoides* sensu Wight, Icon. t. 1946. 1853, non Gardn. 1846.

After studying the above synonymy explain whether the combination *Sarcandra grandifolia* (Miq.) Subr. & Henry was correct or not for the species when placed in the genus *Sarcandra* Gardn.

A. Wight (1853) misidentified this species as *Sarcandra chlorantho-ides* Gardn. *Chloranthus grandifolius* Miq. (1856) was the earliest legitimate name available for this taxon. Hence the later name *Sarcandra irvingbaileyi* Swamy (1953) became superfluous according to Art. 63 of the Code. The combination *Sarcandra grandifolia* (Miq.) Subr. & Henry (1972) was correctly published to retain the legitimate specific epithet "*grandifolius*" in the genus *Sarcandra* Gardn. (Art. 55. 1).

Q. The later homonym *Smithia* Aiton 1789 *nom. cons.* is conserved against *Damapana* Adans. 1763. Whether the earlier homonym *Smithia* Scop. 1777 is legitimate or not ?

A. According to Art. 14.7 of the "Code", a conserved name is conserved against all its earlier homonyms. The generic name

*Smithia* Ait. (1789), conserved against *Damapana* Adans. is thereby conserved automatically against the earlier homonym *Smithia* Scop (1777). Thus the earlier homonym *Smithia* Scop. is made illegitimate by rejection.

## SUGGESTED QUESTIONS

1. Write a brief historical account of the International Code of Botanical Nomenclature.

2. Write short notes on the following:

   (i) ICBN (International Code of Botanical Nomenclature)

   (ii) I A P T (International Association for Plant Taxonomy)

   (iii) Taxon

   (iv) Kew Rule

   (v) Typification

   (vi) Effective and Valid publication

   (vii) Latin diagnosis

3. The following family names do not end in '*aceae*', and not based on the stem of a genus. Are they correct ? What are their alternative names ?

   Palmae, Guttiferae, Leguminosae, Gramineae, Cruciferae, Umbelliferae, Labiatae, Compositae.

4. Define:

   (i) Valid name (ii) Illegitimate name (iii) Correct name (iv) Nomina Generica Conservanda et Rejicienda

5. Define the following with suitable examples:

   (i) Holotype (ii) Paratype (iii) Isotype (iv) Syntype

6. Define :

(i) Lectotype (ii) Neotype

Explain how and when these types are chosen.

7. Explain how the following words/abbreviations are used in botanical literature.

(i) *et* or *et al.* (ii) *ex* (iii) *in* (iv) *emend.* (v) *pro parte* (vi) *sensu amplo* or *sensu lato* (vii) *sensu stricto* (viii) *pro syn.* (ix) *nom. nud.* (x) *auct. non* (xi) *nom. cons.*

8. Write short notes on :

(i) Retention of names or epithets of various taxa

(ii) Rejection of names and epithets

(iii) Choice of names

(iv) Orthography of names and epithets

9. How will you proceed to describe a new genus or species ? Explain.

10. What is a conserved name ? Why conservation is made against the rule of priority ?

11. Explain the following with suitable examples:

(i) Synonym (ii) Homonym (iii) Tautonym (iv) Autonym (v) Basionym (vi) *comb. nov.* (vii) *nom. nov.* (viii) Polynomial (ix) Binomial.

## LITERATURE


Davis, P.H. and Heywood, V.H. *Principles of Angiosperm Taxonomy.* Edinburgh and London, 1963.

Fosberg, F.R. and Sachet, M.H. Manual for Tropical Herbaria. *Reg. Veg.* Vol. 39, Utrecht-Netherlands, 1965.

Heywood, V.H. *Plant Taxonomy.* 2nd edn. London, 1976.

Jain, S.K. and Rao, R.R, *A Handbook of Field and Herbarium Methods.* New Delhi, 1977.

Jeffrey, C. *An Introduction to Plant Taxonomy.* London, 1968.

——— *Biological Nomenclature.* 2nd edn. Great Britain, 1977.

John, H. St. *Nomenclature of Plants.* New York, 1958.

Lawrence, H.M. *Taxonomy of Vascular Plants.* New York, 1951.

McVaugh, R. *et al.* An annotated glossary of Botanical Nomenclature. *Reg. Veg.* Vol. 56. Utrecht-Netherlands, 1968.

Santapau, H. — The International Code of Botanical Nomenclature. *51st. & 52nd Session of Indian Science Congress—Presidential Address*, Calcutta, 1964-65.

———and Henry, A.N. — *A Dictionary of the Flowering Plants in India*. New Delhi, 1973.

Stafleu, F.A. *et al.* — International Code of Botanical Nomenclature. *Reg. Veg.* Vol. 97. Utrecht-Netherlands, 1978.

Stearn, W.T. — *Botanical Latin*. Great Britain, 1966.

24

## GLOSSARY CUM INDEX

| | |
|---|---|
| Affinis, aff. | : Allied to, related to |
| Alternative names | : Two or more names published at the same time by the same author for the same taxon |
| American Code, 5 | |
| Amsterdam Code, 6 | |
| Apud, ap. | : In the work |
| Arbitrary | : Of names formed without regard to etymology or other formal derivation |
| Article | : A series of numbered rules in the Code |
| Auctorum diversorum, auct. div. | : Of various authors |
| Auctorum multorum, auct. mult. | : Of many authors |
| Auctorum non, auct. non, 42 | : Of others, not of |
| Auctorum plurimorum, auct. pl. | : Of most authors |
| Autograph | : A text written by hand |
| Autonym, 21 | : Automatically established name |
| Avowed substitute, 21 | : A name explicitly proposed as a substitute for an existing name |
| Bibliographic error | : Mistake in the number of volume, or page, or other details in citation |
| Binary | : See binomial |

| | |
|---|---|
| Binomial, 10 | : Consisting of a generic name and a specific epithet |
| Brussels Code, 5 | |
| Cambridge Code, 5 | |
| Choice, 44 | |
| Circa, ca. | : About |
| Circumscription | : The diagnostic limit separating the individuals that belong to a taxon from all other individuals |
| Citation, 38 | |
| Combinatio nova, *comb. nov.* 21 | : New combination |
| Correct name, 19 | |
| Cotype | : A term formerly used in botanical nomenclature for syntype and sometimes erroneously for isotype and paratype |
| Determinavit, det. | : Determined |
| Diacritic sign | : An addition to a letter to indicate that its pronunciation is changed |
| Diagnosis, 22 | : See Latin diagnosis |
| Diagnostic character | : Character or feature of a taxon that distinguishes from others |
| Edinburgh Code, 7 | |
| Effective publication, 35 | |
| Emendatus, emend., 39 | |
| Epithet | : Any word in a name of more than one word, subsequent to the generic name |
| Et, 38 | : And |
| Et aliorum, et alii, et al., 38 | : And of others |
| Ex, 39 | : From or according to |
| Exclusus, excl. | : Excluded; used to indicate the excluded elements from the elements included in a former account of a taxon |

Exsiccata : Dried and preserved specimens of plants

Figura, fig., f. : Figure, illustration

Filius : Son

Holotype, 24

Homonym, 20

Ibidem, ibid., ib. : The same

Icone, Ic. : Illustration

Icones inedita, Ic. Inedit. : Inedited plates

Idem quod., i.q. : The same as

Illegitimate, 18

Illustration : A pictorial representation of a plant or parts of a plant

In, 39

In litteris, in litt. : In correspondence

Inclusus, incl. : Included

Indelible : Of written or printed matter that can not be rubbed out

Indirect reference : A reference without complete or precise citation to a previously and effectively published description or diagnosis of a taxon

Ineditus, ined. : Unpublished

Infrageneric : Any taxon of a rank below that of a genus

Infraspecific : Any taxon of a rank below that of a species

International Association for Plant Taxonomy, IAPT, 67

Invalid name : A name which is not validly published

Isotype, 27

Kew Rule, 4

Latin diagnosis, 22

Lectotype, 29

Legitimate, 18

Leningrad Code, 8

Loco cito, loc. cit., l.c. : At the place cited

Manuscriptum, ms.; Manuscripta, mss. : Manuscript; manuscripts *i.e.* Hand-written or typed material existing in only one or few copies

Microfilm : A copy of a published work or manuscript usually much reduced, on a continuous strip of transparent photographic material

Misapplied name, 74 : A name used to refer to a taxon within the range of variation of which its type does not fall

Misidentification, 74 : The assignment of a specimen or specimens to a taxon to which they do not belong

Monotypic : Having only one immediate subordinate taxon ; of a genus with one species or family with one genus

Monstrosity : A plant or specimen showing an abnormal structural condition

Montreal Code, 7

Nec : Not of, nor of

Neotype, 29

Nomen : Name

Nomen aliarum, nom. al. : Names used by other authors

Nomen alternativum, nom. alt. : Alternate name

Nomen ambiguum, nom. ambig. : Name used in different senses

Nomen confusum, nom. conf. : Name causing confusion

Nomen conservandum, nom. cons., 42 &61

Nomen dubium, nom. dub. : Doubtful name

| | |
|---|---|
| Nomen familiarum conservandum, nom. fam. cons., 61 | : Conserved family name |
| Nomen genericum conservandum, nom. gen. cons., 61 | : Conserved generic name |
| Nomen novum, nom. nov., 22 | : New name |
| Nomen nudum, nom. nud., 41 | : Naked name, *i.e.* Name unaccompanied by a description, diagnosis or reference to a published description |
| Nomen obscurum, nom. obsc. | : Obscure name |
| Nomen rejiciendum, nom. rejic. | : A name to be rejected |
| Nomen superfluum, nom. superf. | : See superfluous name |
| Nomen tantum | : Name only |
| Nomen triviale | : Trivial name, specific epithet |
| Nomenclatural synonym, 19 | |
| Nomenclature of taxa, 15 | : Names of taxa |
| Nomenclatural type, 24 | |
| Non | : Not of ; used in the citation of later homonyms to indicate that these refer to taxa other than those described by the authors of the earlier homonyms |
| Non aliorum, non al. | : Not of other authors |
| Non visus, n.v. | : Not seen |
| Nova, nov. | : New |
| Opero citato, op. cit. | : In the work cited |
| Original spelling | : The spelling of a name as it appears in the protologue at the time of its valid publication |
| Orthographic error, 48 | : An unintentional mis-spelling |
| Orthographic variants, 48 | : Alternative spellings of the same name |
| Orthography, 48 | : Spelling |
| Pagina | : Page |
| Paratype, 27 | |
| Paris Code, 4 & 6 | |

| | |
|---|---|
| Parentheses, 40 | : Brackets |
| Polynomial, 10 | |
| Principles, 14 | |
| Priority, 33 | |
| Pro parte, p.p., 40 | : In part |
| Pro parte magna/maxima, p.p. maj./p.p. max. | : For the greater part |
| Pro parte minima, p.p. min. | : For a small part |
| Pro species, pro. sp. | : As a species |
| Pro synonymo, pro syn., 41 | : As a synonym |
| Protologue | : Every thing associated with a name at its first publication, *i.e.* diagnosis, description, illustrations, references, synonymy, geographical data, citation of specimens, discussion and comments |
| Quoad | : As regards; used in citation to indicate what part of a taxon as circumscribed by a previous author is being referred to by the writer |
| Quoad est, q.e. | : Which is |
| Rank, 15 | : The position of a category in the taxonomic hierarchy |
| Recommendations, 14 | |
| Reinstate | : Re-establish |
| Rejected name | : See Nomen rejiciendum |
| Rejection, 46 | |
| Reprints | : Printed matter of the contribution separately from the published regular issue of the main work, and intended for private distribution by the author |
| Resurrect | : Re-establish, Reinstate |
| Retain | : To continue to use it as a correct name |

Retention, 43

Rochester Code, 4

Rules, 14

Seattle Code, 8

Secundum, s.ecus., sec. : According to, following

Sensu amplo, sampl, 40 : In the broad sense

Sensu lato, sens. lat., s.l. : See sensu amplo

Sensu stricto, sens. str., s.s., 40 : In the strict (narrow) sense

Sine numero, s.n. : Without collector's number

Species novum, sp. nov. : New species

Specific epithet : The second part of the binary name of a species

Specimen : A plant or part of a plant, or a number of small plants preserved as a unit for scientific study

Sphalmate, sphalm. : By mistake, in error

Starting points : The dates on which valid publication of name of plants of the different groups are treated as beginning

Status novus, stat. nov., 21 : New status

Stockholm Code, 6

Superfluous name, 46 & 73 : Of a name applied to a taxon for which another name or epithet was already available

Synonymon, syn., 19 : Synonym

Synonymon nova, syn. nov. : Name first used as a synonym

Syntype, 29

Tabula, tab., t. : Plate

Tautonym, 20

Taxon : Defined in the code as a taxonomic group of any rank

Taxonomic synonym, 20

Type method, typification, 23 : The designation or selection of the type of any name

| | |
|---|---|
| Typographic error, 48 | : An error in a printed text for which the printer is responsible not the author |
| Typus conservandus, typ. cons. | : A type to be conserved |
| Unpublished names | : Names that have not been validly published |
| Valid name, 18 | |
| Valid publication, 36 | |
| Varietas nova, var. nov. | : New variety |
| Vienna Code, 5 | |

| | | | |
|---|---|---|---|
| 39 | 4th from top | Chandrab) | Chandrab.). |
| 56 | 2nd from bottom | 6: - 329 | 6: 329 - |
| 64 | 3rd from bottom | ed 1 | ed. 1 |
| 65 | 8th from top | FungiCrteri | Fungi Carteri |
| | 11th from bottom | ot | or |
| 70 | 10th from bottom | nomeclature | nomenclature |
| 71 | 3rd from bottom | Gaertn. Sem. | Gaertn. Fruct. Sem. |
| 72 | 7th from top | ed. (1). | ed. 1). |
| | 20th from top | speces | species |
| | 10th from bottom | lublishing | publishing |
| | 5th from bottom | his | him |
| 74 | 9th from top | indlcated | indicated |
| 75 | 6th from bottom | *Erthroxylum* | *Erythroxylum* |
| 76 | 13th from top | 327.& 604 | 327 & 604 |
| 78 | 3rd from top | *cons*, | *cons*. |
| 80 | 15th from top | flc | fic |
| 81 | 4th from top | speciflc | specific |
| 82 | 4th from bottom | specifle | specific |
| 83 | 13th from top | published by | published by Henry |
| 86 | 2nd from bottom | scop | Scop. |
| 89 | 10th from top | Rao, R.R, | Rao, R.R. |
| 91 | 5th from bottom | Avowed substitute, 21 | Avowed substitute, 22 |
| 96 | 1st from top | Parentheses, 40 | Parentheses, 41 |
| 97 | 5th from top | s. ecus. | secus. |
| | 6th from top | sampl, | s.ampl., |

ERRATA

AN AID TO THE INTERNATIONAL CODE OF BOTANICAL NOMENCLATURE

| Page | Line | For | Read |
|---|---|---|---|
| 3 | 11th from bottom | Linnaeus, | Linnaeus' |
| | 9th from bottom | specifle | specific |
| | 4th from bottom | tainsts | tanists |
| | 3rd from bottom | *Theories* | *Theorie* |
| | 2nd from bottom | dela | de la |
| 5 | 13th from bottom | proposal | proposals |
| | 12th from bottom | *nomtna* | *nomina* |
| | 11th from bottom | requirements | requirements |
| 8 | 17th from top | decṫsions | decisions |
| 10 | 7th from top | *raspathana.* | *raspathana,* |
| 14 | 12th from bottom | effect | affect |
| 15 | 10th from bottom | *Regnum* | *Regnum,* |
| 16 | 13th from top | Lamiacae | Lamiaceae |
| | 14th from top | Papilionacea | Papilionaceae |
| 18 | 8th from top | Thls | This |
| 20 | 4th from top | heteratypic | heterotypic |
| 21 | 2nd from bottom | Borrs. | Borss. |
| 23 | 8th from top | ihe | the |
| | 9th from bottom | thc | the |
| 24 | 17th from top | wheather | whether |
| 29 | 3rd from bottom | selectiag | selecting |
| 30 | 6th from top | [J.] Bombay | [J. Bombay |
| | 6th from top | 72. | 72: |